Averaging in Stability Theory

Mathematics and Its Applications (*Soviet Series*)

Volume 79

Averaging in Stability Theory

A Study of Resonance Multi-Frequency Systems

by

M. M. Hapaev
Department of Computation Mathematics and Cybernetics,
Moscow State University,
Moscow, Russia

KLUWER ACADEMIC PUBLISHERS
DORDRECHT / BOSTON / LONDON

Library of Congress Cataloging-in-Publication Data

```
Khapaev, M. M. (Mikhail Mikhailovich)
   [Usrednenie v teorii ustoichivosti. English]
   Averaging in stability theory : a study of resonance multi
-frequency systems / M.M. Hapaev.
       p.   cm. -- (Mathematics and its applications ; 79)
   Translation of: Usrednenie v teorii ustoichvosti.
   Includes bibliographical references.
   ISBN 0-7923-1581-2 (alk. paper)
   1. Stability.   I. Title.   II. Series.
QA871.K46613   1992
003'.7--dc20                                    91-44079
```

ISBN 0-7923-1581-2

Published by Kluwer Academic Publishers,
P.O. Box 17, 3300 AA Dordrecht, The Netherlands.

Kluwer Academic Publishers incorporates
the publishing programmes of
D. Reidel, Martinus Nijhoff, Dr W. Junk and MTP Press.

Sold and distributed in the U.S.A. and Canada
by Kluwer Academic Publishers,
101 Philip Drive, Norwell, MA 02061, U.S.A.

In all other countries, sold and distributed
by Kluwer Academic Publishers Group,
P.O. Box 322, 3300 AH Dordrecht, The Netherlands.

Translated by V.A. Tkachenko and V.I. Rublinetsky

This book is the updated and revised translation of the original work
Averaging in Stability Theory
© Nauka, Moscow 1986

Printed on acid-free paper

SERIES EDITOR'S PREFACE

'Et moi, ..., si j'avait su comment en revenir,
je n'y serais point allé.'

Jules Verne

The series is divergent; therefore we may be
able to do something with it.

O. Heaviside

One service mathematics has rendered the
human race. It has put common sense back
where it belongs, on the topmost shelf next
to the dusty canister labelled 'discarded non-
sense'.

Eric T. Bell

Mathematics is a tool for thought. A highly necessary tool in a world where both feedback and non-linearities abound. Similarly, all kinds of parts of mathematics serve as tools for other parts and for other sciences.

Applying a simple rewriting rule to the quote on the right above one finds such statements as: 'One service topology has rendered mathematical physics ...'; 'One service logic has rendered computer science ...'; 'One service category theory has rendered mathematics ...'. All arguably true. And all statements obtainable this way form part of the raison d'être of this series.

This series, *Mathematics and Its Applications*, started in 1977. Now that over one hundred volumes have appeared it seems opportune to reexamine its scope. At the time I wrote

"Growing specialization and diversification have brought a host of monographs and textbooks on increasingly specialized topics. However, the 'tree' of knowledge of mathematics and related fields does not grow only by putting forth new branches. It also happens, quite often in fact, that branches which were thought to be completely disparate are suddenly seen to be related. Further, the kind and level of sophistication of mathematics applied in various sciences has changed drastically in recent years: measure theory is used (non-trivially) in regional and theoretical economics; algebraic geometry interacts with physics; the Minkowsky lemma, coding theory and the structure of water meet one another in packing and covering theory; quantum fields, crystal defects and mathematical programming profit from homotopy theory; Lie algebras are relevant to filtering; and prediction and electrical engineering can use Stein spaces. And in addition to this there are such new emerging subdisciplines as 'experimental mathematics', 'CFD', 'completely integrable systems', 'chaos, synergetics and large-scale order', which are almost impossible to fit into the existing classification schemes. They draw upon widely different sections of mathematics."

By and large, all this still applies today. It is still true that at first sight mathematics seems rather fragmented and that to find, see, and exploit the deeper underlying interrelations more effort is needed and so are books that can help mathematicians and scientists do so. Accordingly MIA will continue to try to make such books available.

If anything, the description I gave in 1977 is now an understatement. To the examples of interaction areas one should add string theory where Riemann surfaces, algebraic geometry, modular functions, knots, quantum field theory, Kac-Moody algebras, monstrous moonshine (and more) all come together. And to the examples of things which can be usefully applied let me add the topic 'finite geometry'; a combination of words which sounds like it might not even exist, let alone be applicable. And yet it is being applied: to statistics via designs, to radar/sonar detection arrays (via finite projective planes), and to bus connections of VLSI chips (via difference sets). There seems to be no part of (so-called pure) mathematics that is not in immediate danger of being applied. And, accordingly, the applied mathematician needs to be aware of much more. Besides analysis and numerics, the traditional workhorses, he may need all kinds of combinatorics, algebra, probability, and so on.

In addition, the applied scientist needs to cope increasingly with the nonlinear world and the extra mathematical sophistication that this requires. For that is where the rewards are. Linear models are honest and a bit sad and depressing: proportional efforts and results. It is in the non-linear world that infinitesimal inputs may result in macroscopic outputs (or vice versa). To appreciate what I am hinting at: if electronics were linear we would have no fun with transistors and computers; we would have no TV; in fact you would not be reading these lines.

There is also no safety in ignoring such outlandish things as nonstandard analysis, superspace and anticommuting integration, p-adic and ultrametric space. All three have applications in both electrical engineering and physics. Once, complex numbers were equally outlandish, but they frequently proved the shortest path between 'real' results. Similarly, the first two topics named have already provided a number of 'wormhole' paths. There is no telling where all this is leading - fortunately.

Thus the original scope of the series, which for various (sound) reasons now comprises five sub-series: white (Japan), yellow (China), red (USSR), blue (Eastern Europe), and green (everything else), still applies. It has been enlarged a bit to include books treating of the tools from one subdiscipline which are used in others. Thus the series still aims at books dealing with:

- a central concept which plays an important role in several different mathematical and/or scientific specialization areas;
- new applications of the results and ideas from one area of scientific endeavour into another;
- influences which the results, problems and concepts of one field of enquiry have, and have had, on the development of another.

The shortest path between two truths in the real domain passes through the complex domain.

J. Hadamard

La physique ne nous donne pas seulement l'occasion de résoudre des problèmes ... elle nous fait pressentir la solution.

H. Poincaré

Never lend books, for no one ever returns them; the only books I have in my library are books that other folk have lent me.

Anatole France

The function of an expert is not to be more right than other people, but to be wrong for more sophisticated reasons.

David Butler

Bussum, March 1992

Michiel Hazewinkel

Contents

viii

Preface

Lyapunov's second method is the one generally applied in the analysis of stability. The main difficulty in applying this method lies in constructing the Lyapunov function, especially when the right-hand sides (RHS) of a system contain oscillating terms. That was the reason for the failure of all efforts aimed at the application of Lyapunov's second method to study stability in classical and modern problems of celestial and nonlinear mechanics. A classical problem whose stability in various aspects has been studied for a long time is the three – body problem in celestial mechanics. This problem deals with the motion of two planets around a massive sun, a small parameter in the problem being the ratio of the mass of one of the planets to that of the sun. Without taking into account the interaction between the planets, the trajectory of each planet is an ellipse. In our solar system, Jupiter and Saturn have much greater masses than the other planets; in the three – body problem concerning these planets and the Sun, the small parameter is of the order $10^{-3} - 10^{-4}$. In this system, the eccentricities and the orbit inclinations are small (10^{-1}). If we take into account the interaction between the planets, the motion becomes much more complicated. The planets move near the elliptic orbits, but the parameters of motion undergo small variations. The system is characterized by two frequencies, namely, those of rotations of each planet around the Sun. In the course of investigation of the three – body problem, the canonical variables were introduced and the classical techniques of perturbation theory were developed.

The three – body problem was studied in works of H.Poincaré, where stability was analyzed by methods of perturbation theory. The RHS of the equations were expanded into a fast convergent Fourier series in angular variables, whose coefficients are powers of eccentricity and inclination. The main difficulty in the perturbation theory is caused by small denominators or, in other words, small resonance frequencies. The disappearance of a combinative frequency in a multi – frequency system is called resonance. Since the Hamiltonians of such systems are analytic, there is an infinite number of small resonance frequencies. Beginning with Gauss, researchers have been pointing out that, in problems in celestial mechanics, the RHS of equations contain both fast oscillating terms and slowly varying ones, the latter determining the slow evolution of variables. Concerning the former terms, it has been noted that they only slightly affect the motion, and therefore may be omitted. The process of omitting the fast oscillating terms in RHS was called averaging. Much later, in the works of N.M.Krylov and A.A.Bogolyubov, the investigations started with aim to justify averaging. Thus, it became necessary to study the stability and averaging in problems of celestial mechanics. The main difficulty here is posed by resonance frequencies and small denominators.

Beside the classical three – body problem, the class of resonance problems includes the problem of stability of the satellite motion with respect to the mass centre, the problem of ensuring a stable resonance motions of freely suspended gyrocompas, and some other problems. All of them are described by systems of ordinary differential equations with perturbations; from the viewpoint of stability theory, they pertain to so-called 'critical' or 'neutral' situations, when the properties of the system depend on those of the perturbation.

For such systems, this book presents a generalization of Lyapunov's second method of stability analysis. The generalization concerns both basic assumptions of the method. One of them is the positive – definiteness of the Lyapunov function, the second pertains to the sign of its derivative. The generalization is carried out by combining Lyapunov's second method with the asymptotic averaging method. We introduce a generalized Lyapunov function which is linked with some neighbourhood of the point whose stability is being analyzed. Since the generalized function may be not bounded or even undefined in the equilibrium state, certain restrictions on this state are imposed in an annular region. Certain conditions are also imposed on the sign of the averaged derivative of the generalized Lyapunov function. This averaged derivative is calculated by integration along integral curves of the unperturbed system.

This generalization of Lyapunov second method is intended to be applied in multi – frequency systems with resonance harmonics. We also suggest a new method to estimate small denominators in multi – frequency systems, the method which makes use of the nonlinear properties of frequencies of the system. The generalized Lyapunov function can be constructed for such systems. We will prove theorems on averaging in one – and multi – frequency systems under general assumptions about their RHS. The suggested methods are extended to integro – differential equations, to retarded differential equations, and to partial derivative equations with small nonlinear terms.

The above – mentioned applied problem described by multi – frequency systems will be also considered in this book. In the classical three – body problem, stability on a finite interval will be established.

In the three – body problem we will suggest a hydrodynamical model for planets in which one takes into account the finite dimension and the oblate shape of planets, as well as the constant inclination of the rotation axes of planets. Using the technique of generalized Lyapunov function, we find that this model is stable on the infinite time interval.

CHAPTER 1

Averaging of Ordinary Differential Equations in One – and Multi – Frequency Systems

In this chapter we will prove theorems on averaging under general assumptions on right hand sides (RHS) of a system. These equations will be proved by comparing solutions of the initial and averaged systems. Theorems will also be proved on averaging in standard – form systems, in systems containing slow and fast variables of generic form, and in multi – frequency resonance systems. For a multi – frequency system, we introduce the additional condition that the integral curves of initial system should intersect the resonance line; this condition is being connected with resonance harmonics only.

1.1 Introduction

The important class of nonlinear systems of differential equations consists of systems which admit averaging. Such systems describe either

many oscillatory processes, or the processes close to the periodic ones
with many slowly varying parameters. The systems contain a small parameter of "slowness" which determines the time scale of slow evolution
of parameters of the system.

Averaging procedures which intend to omit fast oscillating terms in
RHS of equations have been used in celestial mechanics for long time,
but justification of averaging methods in generic nonlinear systems was
presented first in [67] and [11], where a concept of the system in standard form was introduced:

$$\frac{dx}{dt} = \mu X(t, x), \quad x(0) = x_0 \tag{1.1}$$

where μ is a small parameter, x and X are n-dimensional vectors.

For the system (1.1), the averaged system takes the form

$$\frac{d\xi}{dt} = \mu X_0(\xi), \quad x(0) = x_0 \tag{1.2}$$

where the RHS in the last equation is obtained by averaging the RHS
of the system (1.1), to wit

$$X_0(x) = \lim_{T \to \infty} T^{-1} \int_0^T X(t, x)\, dt\,. \tag{1.3}$$

The averaged system is substantially simpler than the initial one:
it is autonomous and does not contain a small parameter (it suffices to
make a substitution $\mu t = \tau$); finally it can be integrated numerically or
analytically on the finite interval $\tau \in [0, L]$ with $t \in [0, L/\mu]$.

As the first result, a theorem (the averaging principle) was proved,
stating that the solutions of (1.1) and (1.2) are close to each other on
asymptotically large time interval ($t \in [0, L/\mu]$), if the system (1.1) has
a solution and the RHS of (1.1) satisfies the Lipschitz condition. Later
the averaging principle was extended to integro – differential equations
[28], partial differential equations [89], retarded equations [40], etc. A
survey of results related to averaging is given in [87]. In this chapter we will present new results on averaging based on rather general
assumptions on the system of differential equations to be averaged.

1.2   Averaging in Standard – Form Systems

We will consider systems similar to the ones described by Eq. (1.1), whose RHS will be represented in the form

$$X(t, x) = X_0(t, x) + \tilde{X}(t, x) \tag{2.1}$$

where we singled out the function $\tilde{X}(t, x)$ with the mean value zero; we will also consider the system

$$\dot{\xi} = \mu X_0(t, \xi), \quad \xi(0) = x_0. \tag{2.2}$$

Theorem 1.1 *Let a vector function $X(t, x)$ be defined in the region $P\{t \geq 0, x \in D\}$ and let the following conditions be satisfied:*

1. *$X(t, x)$ satisfies the Caratheodory conditions for existence of continuous solution $x(t)$ with $x(0) = x_0$ ($X(t, x)$ is measurable with respect to t for any fixed x and continuous in x for any fixed t).*

2. *There exists a summable function $M(t)$ and a constant M_0 such that the estimate $\| X(t, x) \| \leq M(t)$ holds in the region P for any finite time interval $[t_1, t_2]$*

$$\int_{t_1}^{t_2} M(t)\, dt \leq M_0(t_2 - t_1).$$

3. *There exists a summable function $H(t)$ and a constant H_0, as well as a non – decreasing function $\Psi(\alpha)$, $\lim_{\alpha \to 0} \Psi(\alpha) = 0$, such that in the region P*

$$\| X(t, x') - X(t, x'') \| < \Psi(|x' - x''|)H(t)$$
$$\int_{t_1}^{t_2} H(t)\, dt \leq H_0(t_2 - t_1)$$

on any finite time interval $[t_1, t_2]$.

4. The limit

$$\lim_{T \to \infty} T^{-1} \int_0^T \tilde{X}(t, x)\, dt = 0$$

uniformly exists for $x \in D$.

5. $X_0(t, x)$ satisfies the Lipschitz condition

$$\| X_0(t, x') - X_0(t, x'') \| < N \| x' - x'' \|$$

in the region P.

6. A solution $\xi = \xi(t)$ of the system (2.2) is defined for all $t \geq 0$ and lies, together with its entire neighbourhood in the domain D.

Then for any $\epsilon > 0$ and $L > 0$ there exists $\mu_0(\epsilon)$ such that for $0 < \mu \leq \mu_0$ and $t \in [0, L/\mu]$

$$\| x(t) - \xi(t) \| \leq \epsilon.$$

Proof. Let us estimate directly deviation of the solution $x(t)$ of the system (2.1) from the solution $\xi(t)$ of the system (2.2) on the interval $t \in (0, L/\mu)$. To do that, we replace the systems (2.1) and (2.2) by the following system of integral equations:

$$x(t) = x_0 + \mu \int_0^t X(t, x(t))\, dt \qquad (2.3)$$

$$\xi(t) = x_0 + \mu \int_0^t X_0(t, \xi(t))\, dt. \qquad (2.4)$$

We subtract (2.4) from (2.3) and then add and subtract $X_0(t, x(t))$ to obtain

$$x(t) - \xi(t) = \mu \int_0^t [X(t, x(t)) - X_0(t, \xi(t))]\, dt +$$

$$+ \mu \int_0^t [X_0(t, x(t)) - X_0(t, x(t))]\, dt.$$

Using condition 5, we find

$$\| x(t) - \xi(t) \| < N\mu \int_0^t \| x(t) - \xi(t) \|\, dt + \mu \left\| \int_0^t \tilde{X}(t, x)\, dt \right\| \quad (2.5)$$

This integral equation estimates the difference $x - \xi$ in the most evident way. Indeed, the magnitude of this difference is determined by the mean deviation of $X(t, x)$ from $X_0(t, x)$ along the integral curve $x(t)$.

Let us estimate the last integral in (2.5) on the interval $[0, L/\mu]$. To this end, we expand the integration over the entire interval of the length L/μ by assuming that $\tilde{X}(t, x)$ equals zero on the right of t and by dividing the interval into n equal parts:

$$\mu \int_0^t \tilde{X}(t, x(t)) \, dt = \mu \sum_{i=0}^{n-1} \int_{t_i}^{t_{i+1}} \tilde{X}(t, x(t)) \, dt =$$

$$= \sum_{i=0}^{n-1} \mu \int_{t_i}^{t_{i+1}} [\tilde{X}(t, x(t)) - \tilde{X}(t, x_i)] \, dt + \tag{2.6}$$

$$+ \sum_{i=0}^{n-1} \mu \int_{t_i}^{t_{i+1}} \tilde{X}(t, x_i) \, dt.$$

Here, $\tilde{X}(t, x_i)$ was added and subtracted from the RHS and x_i is the value taken by the continuous solution $x(t)$ at the division points.

We estimate the first term in the sum in (2.6) by using conditions (b) and (c). According to condition (b), on each segment $[t_j, t_{j+1}]$ we have

$$\| \, x(t) - x_0 \, \| \leq \mu \int_{t_i}^{t_{i+1}} M(t) \, dt \leq \mu \frac{L}{\mu n} M_0 = \frac{L M_0}{n}. \tag{2.7}$$

Conditions (b) and (e) imply that there exist a function $H_1(t)$ and a constant H_1 satisfying condition (b) such that

$$\| \, \mu \int_{t_i}^{t_{i+1}} [\tilde{X}(t, x(t)) - \tilde{X}(t, x_i)] \, dt \, \| \leq$$

$$\tag{2.8}$$

$$\leq \mu \psi \left(\frac{L N_0}{n} \right) \int_{t_i}^{t_{i+1}} H_1(t) \, dt \leq \psi \left(\frac{L N_0}{n} \right) \frac{L H_1}{n}.$$

The first sum in (2.6) contains n such integrals. Hence, for every $\epsilon > 0$ there exists a number n such that

$$\| \sum_{i=0}^{n-1} \mu \int_{t_i}^{t_{i+1}} [\tilde{X}(t, x(t)) - \tilde{X}(t, x_i)] \, dt \, \| \leq \frac{\epsilon}{2} e^{-NL} \tag{2.9}$$

uniformly with respect to μ $(0 < \mu < \mu_1)$.

Let us now fix n and, using the fact that $\tilde{X}(t, x)$ has the zero mean value and let us estimate the second term in (2.6). By virtue of condition (d), there exists a monotonically decreasing function $f(t)$ which tends to zero as $t \to \infty$, such that in the entire domain D

$$\| \int_0^t \tilde{X}(t, x)\, dt \| < tf(t). \tag{2.10}$$

If t belongs to any segment $(t_i, t_{i+1}]$, except the first one, then for every t_i

$$\| \mu \int_0^{t_i} \tilde{X}(t, x_i)\, dt \| < \frac{\mu L}{\mu} f\left(\frac{L}{\mu n}\right) = F_1(\mu) \tag{2.11}$$

where $F_1(\mu)$ tends to zero as $\mu \to 0$.

For the first integral in the considered sum we have

$$\| \mu \int_0^t \tilde{X}(t, x_0)\, dt \| \leq \mu t f(t) \leq F_2(\mu) \tag{2.12}$$

where $F_2(\mu) = \sup \| \tau f\left(\frac{\tau}{\mu}\right) \|$, $\tau = \mu t$, $\tau \in [0, L]$, $\tau \leq \frac{L}{n}$, $\lim_{\mu \to 0} F_2(\mu) = 0$,

$$n \| \mu \int_{t_i}^{t_{i+1}} \tilde{X}(t, x_i)\, dt \| < \tag{2.13}$$

$$< n\mu \| \int_0^{t_{i+1}} \tilde{X}(t, x_i)\, dt \| + n\mu \| \int_0^{t_i} \tilde{X}(t, x_i)\, dt \|$$

whence

$$\| \sum_{i=0}^{n-1} \mu \int_{t_i}^{t_{i+1}} \tilde{X}(t, x_i)\, dt \| < 2nF_1(\mu) + F_2(\mu). \tag{2.14}$$

If n is fixed, then $F_1(\mu)$ and $F_2(\mu)$ tend to 0, as $\mu \to 0$, so that for any $\epsilon > 0$ there exists $\mu_0 < \mu_1$ such that for every μ $(0 < \mu < \mu_0)$ we get

$$\| \sum_{i=0}^{n-1} \mu \int_{t_i}^{t_{i+1}} \tilde{X}(t, x_i)\, dt \| < \frac{\epsilon}{2} e^{-NL}. \tag{2.15}$$

Combining (2.9) and (2.15), we obtain the estimate for $t \in [0, L/\mu]$:

$$\| \mu \int_0^t \tilde{X}(t, x(t))\, dt \| < \epsilon e^{-NL}. \tag{2.16}$$

The inequalities (2.16), (2.15) and the Gronwall lemma [99] imply the assertion of the theorem:

$$\| x(t) - \xi(t) \| < \epsilon.$$

The proof given above was first published in Ref. [54], where, for the case $X_0(t, x) = X_0(x)$, i.e. when the non – averaged part is independent of time, the following theorem was proved.

Theorem 1.2 *Suppose that conditions (a) to (f) of Theorem 1.1 are satisfied and that there exists the limit*

$$\lim_{T \to \infty} \frac{1}{T} \int_0^T X(t, x) \, dt = X_0(t)$$

uniform with respect to $x \in D$. Then the assertion of Theorem 1.1 is true.

Remark. In proving Theorem 1.1 one could use instead of the Caratheodory existence theorem a more general theorem (e.g. that of Ref. [31]). Thus, the theorems formulated above extend the averaging principle to the systems with discontinuous and unbounded RHS.

It is pointed out in Ref. [30] that one can average not the whole RHS of a system, leaving the time dependence in the RHS of the simplified system. The corresponding averaging schemes are justified by Theorem 1.1.

1.3 Averaging in Systems with Slow and Fast Variables

Let us consider a system with slow variables x (dim $x = n$, $x \in D \subset \mathbf{R}^n$) and fast variables y (dim $y = m$, $y \in Q \subset \mathbf{R}^m$), $0 < \mu \le \mu_0$;

$$\dot{x} = \mu X(x, y, t, \mu), \quad \dot{y} = \mu Y(x, y, t, \mu). \tag{3.1}$$

The averaging procedure for system similar to (3.1) was justified in Ref. [22]. The system (3.1) can be reduced to a standard – form system, provided that the integrals of the degenerate system

$$\dot{y} = Y(x, y, t, 0), \quad x = \text{const} = x_0 \tag{3.2}$$

are known and are taken as additional slow variables.

In what follows we are going to formulate and prove a theorem on averaging the system (3.1) under general assumptions about RHS of this system.

We assume that there exists the limit

$$\lim_{T \to \infty} \frac{1}{T} \int_{t_0}^{t_0+T} X(x_0, y(t, 0), t, 0)\, dt = X_0(x_0) \tag{3.3}$$

uniform with respect to the initial condition $x_0 \in D$, $y_0 \in Q$, $t_0 \geq 0$.

Consider yet another system

$$\dot{\xi} = \mu X_0(\xi). \tag{3.4}$$

Theorem 1.3 *Let the summable functions $M(t)$, $H(t)$, the constants M_0, H_0, and non – decreasing function $\psi_i(\alpha)$, $i = 1, 2, \ldots, 5$, $\lim_{\alpha \to \infty} \psi_i(\alpha) = 0$ exist and satisfy*

1.

$$\| X(x'', y'', t, \mu) - X(x', y', t, 0) \| \leq$$
$$\leq H(t)[\psi_1(\| x'' - x' \|) + \psi_2(\| y'' - y' \|) + \psi_3(\mu)].$$

2.

$$\| Y(x'', y'', t, \mu) - Y(x', y', t, 0) \| \leq$$
$$\leq H(t)[\psi_4(\| x'' - x' \|) + \| y_2 - y_1 \|) + \psi_5(\mu)].$$

3.

$$\| X(x,y,t,\mu) \| \le M(t)$$

$$\int_{t_1}^{t_2} M(t) \le M_0(t_2 - t_1)$$

$$\int_{t_1}^{t_2} H(t) \le H_0(t_2 - t_1)$$

for any $x \in D$, $y \in Q$, $t \ge 0$, $t_2 \ge t_1 > 0$, $0 < \mu \le \mu_0$.

4. Let the limit (3.4) exist uniformly with respect to the initial conditions.

5. Let the vector function $X_0(x,y,t,\mu)$ satisfy the Lipschitz condition

$$\| X_0(x',y,t,\mu) - X_0(x'',y,t,\mu) \| \le N \| x' - x'' \|$$

in the variable x.

Then, for each $\epsilon > 0$ there exist $L > 0$ and $\mu_0 > 0$ such that if the solution $\xi = \xi(t,\mu)$ of the system (3.4) together with its σ – neighbourhood lies in the domain D for $t_0 \le t \le t_0 + L/\mu$, then $\| x(t,\mu) - \xi(t,\mu) \| < \epsilon$ for all $t_0 \le t \le t_0 + L/\mu$, $0 < \mu \le \mu_0$.

Proof. Let us introduce the notation: $z = (x,y)$, $z_0 = (x_0,y_0)$, $z = z(t,z_0,t_0,\mu)$ being a solution of the system (3.1). Using the integral equations for the systems (3.1), (3.4), we obtain

$$x(t,\mu) - \xi(t,\mu) = \mu \int_0^t [X_0(x(\tau,\mu)) - X_0(\xi(\tau,\mu))]\, d\tau +$$

$$(3.5)$$

$$+\mu \int_0^t [X(z(\tau,\mu),\tau,\mu) - X_0(x(\tau,\mu))]\, d\tau.$$

Taking condition (d) of the theorem into account, we get

$$\| x(t,\mu) - \xi(t,\mu) \| \le \mu N \int_0^t \| x(t,\mu) - \xi(t,\mu) \|\, d\tau +$$

$$(3.6)$$

$$\mu \int_0^t [X(z,\tau,\mu) - X_0(x(\tau,\mu))]\, d\tau.$$

Combining the Gronwall lemma and condition (b), for every z_0, t_0 we can easily estimate the difference

$$\| y(t, z_0, t_0, \mu) - y(t, z_0, t_0, 0) \| \leq \beta(t, \mu) \tag{3.7}$$

where $\beta(t, \mu)$ is a non – decreasing function of t such that $\lim_{\mu \to 0} \beta(t, \mu) = 0$. This estimate was obtained in Ref. [62] (p. 135). Let us denote by $t^*(\mu, k)$ the root of the equation $\beta(t, \mu) = k$, $k > 0$ (if it exists). If this equations does not have solutions, we set $t^* = +\infty$. Let us introduce the function $\Delta(\mu, k) = \min\{\mu^{-\frac{1}{2}}, t^*(\mu, k)\}$. For fixed $k > 0$,

$$\lim_{\mu \to 0} \Delta(\mu, k) = +\infty \tag{3.8}$$

Let us divide the half – axis $t \geq t_0$ into segments of the length Δ by the points $t + j\Delta$. We fix an arbitrary t from the interval $[t_0, t_0 + L\mu]$; then, for any $k = \max_{t_j \leq t}\{j\}$

$$\mu k \leq \Delta^{-1}. \tag{3.9}$$

By virtue of the triangle inequality

$$\begin{aligned}
\| x(t, \mu) - \xi(t, \mu) \| &\leq \\
&\leq \| x(t_k, \mu) - \xi(t_k, \mu) \| + \| x(t, \mu) - x(t_k, \mu) \| + \\
&\quad + \| \xi(t_k, \mu) - \xi(t, \mu) \|
\end{aligned} \tag{3.10}$$

and of the estimates $\| X_0 \| < M_0$, $t - t_k < \Delta$, we get

$$\| x(t, \mu) - \xi(t, \mu) \| \leq \| x(t_k, \mu) - \xi(t_k, \mu) \| + 2M_0\mu\Delta. \tag{3.11}$$

According to (3.10), it is sufficient to estimate the second integral on the RHS of the inequality (3.5) at $t = t_k$

$$\begin{aligned}
\mu \int_{t_0}^{t_k} [X(z, \tau, \mu) - X_0(x)] \, d\tau &= \\
&= \sum_{j=0}^{k-1} \mu \int_{t_j}^{t_{j+1}} [X(z(\tau, z_j, t_j, 0), \tau, 0) - X_0(x_j)] \, d\tau +
\end{aligned}$$

$$+ \sum_{j=0}^{k-1} \mu \int_{t_j}^{t_{j+1}} [X(z(\tau, z_0, t_0, \mu), \tau, \mu) - X(z(\tau, z_j, t_j, 0), \tau, 0))] \, d\tau \, +$$

$$+ \sum_{j=0}^{k-1} \mu \int_{t_j}^{t_{j+1}} [X_0(x_j) - X_0(x(\tau, z_0, t_0, \mu))] \, d\tau. \tag{3.12}$$

Given $\epsilon > 0$ $(\epsilon < \sigma)$, we will choose $k > 0$ to ensure the inequality

$$H_0 \psi_2(k) \leq \frac{\epsilon}{4} e^{-LN}. \tag{3.13}$$

Let us, in turn, estimate the terms on the RHS of Eq. (3.13). According to the condition (3.8) and existence of the uniform mean X_0, it follows that there exist the numbers $T(\epsilon) > 0$ and $\mu_1 > 0$ such that for all μ, $0 < \mu \leq \mu_1$, both the inequality

$$\Delta(\mu, k) \geq T(\epsilon)$$

and the estimate

$$\sum_{j=0}^{k-1} \mu \left\| \int_{t_j}^{t_{j+1}} [X(z(\tau, z_j, t_j, 0), \tau, 0) - X_0(x_j)] \, d\tau \right\| \leq$$

$$\tag{3.14}$$

$$\leq \mu k \Delta \left(\frac{\epsilon}{4}\right) e^{-LN} \leq \left(\frac{\epsilon}{4}\right) e^{-LN}$$

hold. In order to estimate the second sum in (3.12), we will use condition (a) of the theorem together with the inequality $\mu k \Delta \leq 1$, which follows from (3.10) and was already employed above. We find that

$$\sum_{j=0}^{k-1} \mu \left\| \int_{t_j}^{t_{j+1}} [X(z(\tau, z_0, t_0, \mu), \tau, \mu) - X(z(\tau, z_j, t_j, 0), \tau, 0))] \, d\tau \right\| \leq$$

$$\tag{3.15}$$

$$\leq H_0 [\psi_1(\mu k_0 \Delta) + \psi_2(k) + \psi_3(\mu)].$$

Likewise

$$\sum_{j=0}^{k-1} \mu \left\| \int_{t_j}^{t_{j+1}} [X_0(x_j) - X_0(x(\tau, z_0, t_0, \mu))] \, d\tau \right\| \leq \frac{\mu \Delta N M_0}{2}. \tag{3.16}$$

Because of (3.13) – (3.16), Eq. (3.12) implies that

$$\mu \parallel \int_{t_0}^{t_k} [X(z, \tau, \mu) - X_0(x)] \, d\tau \parallel \leq$$

$$\frac{\epsilon}{2} e^{-LN} + H_0[\psi_1(\mu M_0 \Delta) + \psi_3(\mu)] + \frac{\mu \Delta N M_0}{2}. \tag{3.17}$$

Since $\mu\Delta \leq \sqrt{\mu}$, there exists a number $\mu_2 > 0$ such that for all μ, $0 < \mu \leq \mu_2$, the inequality

$$H_0[\psi_1(\mu M_0 \Delta) + \psi_3(\mu)] + \frac{\mu \Delta N M_0}{2} \leq \frac{\epsilon}{2} e^{-LN} \tag{3.18}$$

holds. Combining the last two estimates, we get

$$\mu \parallel \int_{t_0}^{t_k} [X(z, \tau, \mu) - X_0(x)] \, d\tau \parallel \leq \frac{3}{4} \epsilon e^{-LN}. \tag{3.19}$$

Applying the Gronwall lemma to the inequality (3.6) and taking (3.20) into account, we write

$$\parallel x(t_k, \mu) - \xi(t_k, \mu) \parallel \leq \frac{3}{4} \epsilon. \tag{3.20}$$

We now choose μ_3 such that

$$2M_0 \mu \Delta \leq \frac{\epsilon}{4}. \tag{3.21}$$

for $0 < \mu \leq \mu_3$. Combining the inequalities (3.11), (3.20), and (3.21) for $t_0 \leq t \leq t_0 + L/\mu$, we get the desired estimate

$$\parallel x(t, z_0, t_0, \mu) - \xi(t\mu, x_0, t_0) \parallel \leq \epsilon \tag{3.22}$$

where the parameter μ belongs to the interval $0 < \mu \leq \mu_0 = \min\{\mu_1, \mu_2, \mu_3\}$.

1.4   Averaging in Multi – Frequency Systems

A multi – frequency system is a system of ordinary differential equations of the form

$$\dot{x} = \mu X(x, \psi), \quad \dim x = n$$

$$\dot{\psi} = \omega(x) + \mu \Phi(x, \psi), \quad \dim \psi = m \geq 2$$

(4.1)

where $\dot{x} = \dfrac{dx}{dt}$, μ is a small parameter, $x \in D$. The vector – valued functions $X(x, \psi)$ and $\Phi(x, \psi)$ are periodic in ψ variables with the period 2π. The RHS of the system (4.1) satisfy the Lipschitz condition in variables $x \in D$ uniformly with respect to $\psi \in [0, 2\pi]$ and are bounded. We assume also that both X and Φ can be expanded into uniformly and absolutely convergent Fourier series.

We construct the averaged system corresponding to the system (4.1), by formally averaging over angles:

$$\dot{\xi} = \mu X_0(\xi),$$

$$X_0(\xi) = \frac{1}{(2\pi)^m} \int_0^{2\pi} X(x, \psi) \, d\psi.$$

(4.2)

A solution $\xi(\mu t)$ of this system is assumed to be defined, together with its ρ – neighbourhood, in the domain D, for $0 < t < \infty$.

In the one – frequency case, the solutions of the systems (4.1) and (4.2) remain close to each other on an asymptotically long segment, if rather general conditions imposed on the RHS (Theorems 1.1 and 1.2) are satisfied.

In multi – frequency systems similar to (4.1) the pattern of integral curves becomes more sophisticated due to appearance of slowly varying harmonics of the Fourier series $X_k(x) \exp(ik\psi)$ for which the combinative frequencies become small on the RHS. To justify proximity of solutions of (4.1) and (4.2), we will need a condition restricting these harmonics. The indicated closeness of solutions may not be kept, had such restrictions been not imposed. An example of such situation was described in Ref. [7], where the applicability conditions for the averaging techniques were introduced for two – frequency systems with analytic RHS. The main restriction in this paper was related to all harmonics of the function $X(x, \psi)$.

Let us introduce a restriction on the resonance harmonics of the function $X(x, \psi)$ only. To this end, we fix $\epsilon > 0$ and examine the integral curve $\xi = \xi(\mu t)$ of the system (4.2), together with its ϵ – neighbourhood in the domain D. Let us denote by $X_\epsilon(x, \psi)$ the sum of those terms in the Fourier series of $X(x, \psi)$, whose combinative frequencies $k\omega(x)$ become smaller than a certain quantity $\sigma(\epsilon)$ $(\mu \ll \sigma(\epsilon))$ which will be chosen later, in the ϵ – neighbourhood of the curve $\xi = \xi(t)$; these frequencies are called the *resonance* frequencies. Thus,

$$X(x, \phi) = X_0(x) + X_\epsilon(x, \psi) + \tilde{X}(x, \phi) = X_0 + \hat{X} \qquad (4.3)$$

where $\tilde{X}(x, \psi)$ are the terms which oscillate with frequencies not smaller than $\sigma(\epsilon)$.

We consider the derivative of the resonance frequency along the vector

$$X_0(x) + X_\epsilon(x, \psi) = \bar{X}(x, \psi) \qquad (4.4)$$

and calculate the minimum of the modulus of it (we will denote this modulus by a), over all labels k present in $X_\epsilon(x, \psi)$, $\psi \in [0, 2\pi]$, and x along the curve $\xi(\mu t)$, i.e.

$$a = \min \left| \left(\frac{\partial(k\omega(x))}{\partial x} \right) (X_0(x) + X_\epsilon(x, \psi)) \right|. \qquad (4.5)$$

The main restriction is that $a \neq 0$. This means that the derivative of the vector $\omega(x)$ on the curve $\xi = \xi(\mu t)$ calculated along the vector $X(x, \psi)$ is not orthogonal to the resonance value k, for which $k\omega(x)$ is small or equal to zero. One can easily see that even for three – frequency system this condition may be not satisfied if all possible vectors k are present in the RHS, since there can exist a vector k orthogonal to both $\omega(x)$ and its derivative along X. Thus, for this vector the frequency $k\omega$ can remain small during motion.

Now we formulate a theorem on averaging in multi – frequency systems of the type (4.1).

Theorem 1.4 *Let vector functions $X(x, \psi)$ and $\Phi(x, \psi)$ be 2π – periodic in the variable ψ and can be expanded into uniformly and absolutely convergent multiple Fourier series; also, let these vector functions satisfy the Lipschitz conditions with respect to the variables $x \in D$ and be bounded. Let $a > 0$, a being defined by (4.5). Then for every $\epsilon > 0$ there exists μ_0 such that*

$$\| x(t) - \xi(\mu t) \| < \epsilon$$

for $0 < \mu \le \mu_0$ and $0 < t < L/\mu$, where L is a fixed number.

Proof. Let us estimate from above the time interval Δt_σ during which the frequency $k\omega(x)$, being small or equal to zero in the ϵ – neighbourhood of the curve $\xi = \xi(\mu t)$ reach the value $\sigma(\epsilon)$.

The equation which determines $k\omega(x)$ is

$$\frac{d(k\omega)}{dt} = \frac{\partial(k\omega)}{\partial x}[\bar{X}(x, \psi) + \tilde{X}(x, \psi)]. \tag{4.6}$$

According to Theorem 1.1, the solution of this equation is close to the solution of the equation

$$\frac{d(k\omega(x))}{dt} = \frac{\partial(k\omega(x))}{\partial x}\bar{X}(x, \psi) \tag{4.7}$$

since $\tilde{X}(x, \psi)$ does not contain any resonance exponents.

Using the estimate (4.5), we get

$$\Delta t_\sigma < \frac{\sigma}{a\mu}. \tag{4.8}$$

To prove the proximity of solutions of (4.1) and (4.2), we write down the integral equations for these systems and, by subtracting them, we obtain

$$x - \xi = \mu \int_0^t [X_0(x) - X_0(\xi)] \, dt +$$

$$+ \mu \int_0^t [X(x, \psi) - X_0(x)] \, dt. \tag{4.9}$$

Taking the Lipschitz condition for $X_0(x)$ into account, and using the notations of (4.3), we arrive at the inequality

$$\| x - \xi \| < \lambda \int_0^T \| x - \xi \| \, d\tau + \mu \int_0^t \hat{X}(x, \psi) \, dt. \tag{4.10}$$

The terms of the series $\hat{X}(x, \psi)$, whose sum will be denoted by $\tilde{X}_\sigma(x, \psi)$, oscillate with frequencies not less than σ. When at least one of the frequencies becomes less than σ, we exclude this term from $\tilde{X}_\sigma$ and, on the corresponding time interval, include it to $\bar{X}_\sigma(x, \psi)$. Let us represent X in the following manner

$$\hat{X}(x, \psi) = \tilde{X}_\sigma(x, \psi) + \tilde{\bar{X}}_\sigma(x, \psi) + \bar{X}_\sigma(x, \psi) - \tilde{\bar{X}}_\sigma(x, \psi) \tag{4.11}$$

where the terms of the series $\bar{X}_\sigma(x, \psi)$ and $\tilde{\bar{X}}_\sigma(x, \psi)$ differ from zero only on the time intervals $2\Delta T_\sigma$ which satisfy the inequality (4.1). The terms of $\tilde{\bar{X}}_\sigma(x, \psi)$ oscillate on time intervals with fixed frequencies σ. The terms $\tilde{\bar{X}}_\sigma(x, \psi)$ are introduced in order to supplement $\bar{X}_\sigma(x, \psi)$ with terms which oscillate everywhere. We estimate the modulus of the difference $\bar{X}_\sigma - \tilde{\bar{X}}_\sigma$ as follows

$$\mu \left\| \int_0^t [\bar{X}_\sigma - \tilde{\bar{X}}_\sigma] \, dt \right\| < 4 \frac{\sigma(\epsilon)}{a} \sum_k \max_{x \in D} |X_k(x)|. \tag{4.12}$$

Now we take $\sigma(\epsilon)$ small enough to ensure the inequality

$$\mu \left\| \int_0^t [\bar{X}_\sigma - \tilde{\bar{X}}_\sigma] \, dt \right\| < \frac{\epsilon}{2} e^{-\lambda L} \tag{4.13}$$

with $L \geq 0$.

In order to estimate $\mu \int_0^t \tilde{X}_\sigma(x, \psi) \, dt$ on the time interval L/μ, we extend the integration to the entire integral, assuming that $\tilde{X}_\sigma \equiv 0$ on the right of t and divide the interval $[0, L/\mu]$ into subintervals Δt_i, $l \leq \Delta t_i \leq 2l$ (l will be chosen later):

$$\mu \int_0^{L/\mu} [\tilde{X}_\sigma(x, \psi) + \tilde{\tilde{X}}_\sigma(x, \psi)] \, dt =$$

$$= \mu \sum_i \int_{t_i}^{t_{i+1}} [\tilde{X}_\sigma(x, \psi) + \tilde{\tilde{X}}_\sigma(x, \psi) - \tilde{X}_\sigma(x_i, \psi) - \tilde{\tilde{X}}_\sigma(x_i, \psi)] \, dt +$$

$$+ \mu \sum_i \int_{t_i}^{t_{i+1}} [\tilde{X}_\sigma(x_i, \psi) + \tilde{\tilde{X}}_\sigma(x_i, \psi)] \, dt. \qquad (4.14)$$

Here $\tilde{X}_\sigma(x_i, \psi)$ was added to and subtracted from RHS, x_i being the values of the continuous solution x_t at the division points. Since $\tilde{X}_\sigma + \tilde{\tilde{X}}_\sigma$ has zero mean value, there exists a function $f(t)$, monotonically decreasing to zero, such that

$$\| \int_{t_i}^{t_{i+1}} [\tilde{X}_\sigma(x_i, \psi) - \tilde{\tilde{X}}_\sigma(x_i, \psi)] \, dt \leq \Delta t_i f(\Delta t_i). \qquad (4.15)$$

Now we chose the segments Δt_i and the number l to ensure the inequality

$$\mu \| \sum_i \int_{t_i}^{t_{i+1}} [\tilde{X}_\sigma(x_i, \psi) + \tilde{\tilde{X}}_\sigma(x_i, \psi)] \, dt \| <$$

$$\qquad (4.16)$$

$$< \mu f(l) \sum_i \Delta t_i \leq L f(l) < \frac{\epsilon}{3} e^{-\lambda L}.$$

Since $\| X(x, \psi) \| < M$ for $x \in D$,

$$\| x(t) - x_i \| \leq \mu \int_{t_i}^{t_{i+1}} \| X(x_i, \psi) \| \, dt \leq \mu \, 2l M_0. \qquad (4.17)$$

Combining (4.17) end the Lipschitz condition for $\tilde{X}(x, \psi)$ and $\tilde{\tilde{X}}_\sigma(x_i, \psi)$ with the Lipschitz constant N, we get

$$\sum_i \mu \int_{t_i}^{t_{i+1}} \| [\tilde{X}_\sigma(x, \psi) + \tilde{\tilde{X}}_\sigma(x, \psi)] -$$

$$-[\tilde{X}_\sigma(x,\psi) + \bar{\tilde{X}}_\sigma(x,\psi)] \parallel dt \le \qquad (4.18)$$
$$\le \mu\, 2lMNL < \frac{\epsilon}{3} e^{-\lambda L};$$

the last inequality holds if $\mu < \mu_0$, with $\mu_0(\epsilon)$ being sufficiently small.

Combining the estimates (4.13), (4.16), and (4.18) for $\mu < \mu_0(\epsilon)$ and $0 < t < L/\mu$, we obtain

$$\parallel \mu \int_0^t \hat{X}(x,\psi)\, dt \parallel < \epsilon e^{-\lambda L}. \qquad (4.19)$$

The inequalities (4.19) and (4.10) together with the Gronwall lemma imply that

$$\parallel x(t) - \xi(\mu t) \parallel < \epsilon \quad 0 < t < \frac{L}{\mu}. \qquad (4.20)$$

The theorem which we have just proved was first published in [55].

As was stated earlier, the main condition. characteristic for the multi – frequency systems of the form (4.5) means that the vector of the averaged system with its resonance harmonics does not lie on the resonance curve. Moreover, the magnitude of the oscillating harmonics is not significant and the solution of the averaged system 'brings' the solution of the complete system.

Another approach to this problem is possible if certain harmonics are included into the averaged system. It is, in particular, reasonable to include into the averaged system the slow oscillating resonance harmonics. We will refer to such a system as to an enlarged averaged system.

The proof of the proximity of the solutions of the complete and enlarged averaged systems is similar to that of Theorem 1.4, with alterations specific to the proof of Theorem 1.1. An enlarged averaged system is not autonomous, but its numerical integration is much easier compared to the initial one, since it does not contain rapidly oscillating harmonics, whose presence forces us to choose a small integration step.

CHAPTER 2

Generalization of Lyapunov Second Method and Averaging in Stability Theory

In the present chapter we propose a generalization of Lyapunov second method, the generalization which is oriented at multi − frequency systems and is based on a combination of Lyapunov second method and the method of asymptotic averaging. The generalization is made in two directions corresponding to two limitations of Lyapunov second method. A system with perturbations is considered. It is assumed that a system without perturbations has a stable equilibrium position; it is assumed also that there exists a Lyapunov function v_0 of a not perturbed system, having a non − positive derivative. the system with perturbations is assumed to have a perturbed (generalized) Lyapunov function $v = v_0 + u$, such that in the annular region containing the equilibrium position, the perturbation can be made arbitrary small by choosing a sufficiently small u. Theorems on stability on infinite and asymptotically large intervals will be proved. Some systems will be examined under additional assumptions which simplify application of theorems of the generalized Lyapunov method.

2.1   Introduction

In this chapter we study systems of ordinary differential equations of the form

$$\frac{\mathrm{d}z}{\mathrm{d}t} = f(t, z) + \mu R(t, z, \mu) \quad z = (x, y) \tag{1.1}$$

$\dim x = n$, $\dim y = m$; μ is a small parameter, $0 < \mu \ll 1$. The perturbations $\mu R(t, z, \mu)$ can be expanded in powers of μ. The system (1.1) will be considered in the domain P:

$$P = P\{t \geq 0, \parallel x \parallel < H, \ y \in D\} \tag{1.2}$$

where D is the variational range of y. We will assume that the function $f(t, z)$ satisfies the Lipschitz condition with the Lipschitz constant N with respect to z in the domain D. Let $f(t, z)$ and $R(t, z, \mu)$ for $0 < \mu < \mu_0$ satisfy also the conditions ensuring existence and uniqueness of a solution of the problem with the initial data in the domain P.

Moreover, we assume that the non – perturbed system

$$\frac{\mathrm{d}z}{\mathrm{d}t} = f(t, z) \tag{1.3}$$

has a stationary point in x, i.e. $f(t, 0, y) \equiv 0$ for $1 \leq i \leq n$. the perturbed system will be examined for its stability in x in the neighbourhood of the point $x = 0$.

It is known, cf. [75] that for perturbations which are small in a certain sense, Lyapunov second method makes it possible to determine the stability of the stationary point, provided that this point is an asymptotically stable configuration of equilibrium of the system (1.3). For such systems, theorems on stability under constantly acting perturbations hold.

In the past, theorems were proved (cf. [34], [66]) according to which, if the equilibrium position is asymptotically stable and uniform with respect to the initial data, then the solutions are stable in the presence of permanently acting perturbations, being in average bounded. It was also proved that presence of sufficiently fast oscillations in a uniformly asymptotically stable system, even in a case when the amplitude of such oscillations is not very small, do not impair the stability in the significant way. Further weakening of restrictions on the unperturbed system (1.3) is due to introduction of some additional information about the perturbations, or on small forces μR acting on the system.

The present chapter is devoted to studying the system (1.1) in the so – called 'neutral' case, when the system (1.3) has a stable equilibrium position only. In this case, the stability or instability of the system (1.1) will be determined by the properties of the μR forces. The small forces μR, acting on a stable system, can either destroy the stability of the equilibrium position, or preserve it, or even create asymptotic stability. Such properties of the system are described by the generalized Lyapunov method, moreover the standard requirements concerning the Lyapunov function and its derivatives are replaced by much less restrictive ones.

Let us assume that the point $x = 0$ is a stable stationary point of the unperturbed system. Let the stability of this point be ensured by existence of a positive – definite Lyapunov function $v_0(t, x, y)$ which admits an infinitely small upper bound in x. Let us assume also that the derivative of y_0 calculated from Eq. (1.3) is non – positive. The latter means that for the perturbed system (1.1) we have a 'neutral case' of stability.

Generalizations of Lyapunov second method will be carried out in two directions: firstly, we will weaken the condition for the Lyapunov function to be positive – definite and, secondly, we will weaken the limitation concerning the sign of derivative.

For the perturbed system (1.1), we will construct the perturbed Lyapunov function

$$v = v_0(t, x, y) + u(t, x, y, \mu, \epsilon) \tag{1.4}$$

where the perturbation u, which positive definiteness is not assumed, can be chosen small for small μ; ϵ is the size of the neighbourhood of the point whose stability is being investigated. Thus, the condition for the Lyapunov function to be positive – definite is generalized, since in the neighbourhood of the equilibrium position $x = 0$ the function v is not positive – definite. Let us differentiate the perturbed Lyapunov function v and use (1.1)

$$
\frac{dv}{dt} = = \frac{\partial v_0}{\partial t} + \frac{\partial v_0}{\partial z} f + \frac{\partial v_0}{\partial z} \mu R +
$$
$$
+ \frac{\partial u}{\partial t} + \frac{\partial u}{\partial z} f + \frac{\partial u}{\partial z} \mu R. \tag{1.5}
$$

On the RHS of this expression we consider the terms containing R and u, single out the factor $\theta(\mu)$ depending on μ, and introduce the notation

$$\theta(\mu)\phi(t, z, \mu) = \frac{\partial v_0}{\partial z}\mu R + \frac{\partial u}{\partial t} + \frac{\partial u}{\partial z} f + \frac{\partial u}{\partial z}\mu R \tag{1.6}$$

to ensure the condition $\phi(t, z, 0, \epsilon) \not\equiv 0$. Let us consider the integral

$$J(t_0, z_0, T, \mu, \epsilon) = \int_{t_0}^{t_0+T} \phi(t, \bar{z}(t), \mu, \epsilon)\, dt \tag{1.7}$$

which is calculated by integrating along the integral curves $z = z(t)$ of the unperturbed system (1.3), with the initial condition $z(t_0) = z_0$, or let us consider the mean $\psi(t_0, z_0, \mu, \epsilon)$ (if it exists)

$$\psi(t_0, z_0, \mu, \epsilon) = \lim_{T \to \infty} \frac{1}{T} J(t_0, z_0, T, \mu, \epsilon). \tag{1.8}$$

The second condition of the Lyapunov method which is the assumption that the derivative of the Lyapunov function is non – positive, can be generalized as follows: we require only the negativeness of ψ (the mean of the derivative of the generalized Lyapunov function), or of the integral J.

The stability, which is ensured by existence of the perturbed Lyapunov function (1.4) and restrictions on the sign of the mean (1.4), consists of the following: for each $\epsilon > 0$ there exist $\eta > 0$ and μ_0 such that the integral curves of the point $x = 0$ remain within the ϵ – neighbourhood for all $t > 0$, provided that the parameter μ is sufficiently small.

It should be noted that the above – defined stability coincides with the Lyapunov stability with respect to the variables x, μ for the enlarged system, which is the result of adding the equation $\dot{\mu} = 0$ to the system (1.1). The equilibrium position stable in this sense will be called the stable position with respect to some variables and the parameter or, in other words, (x, μ) – stable.

From the practical point of view, the case when nothing is known about the sign of the perturbed Lyapunov function $\phi(t, z, \mu, \epsilon)$, the integral J, and about the mean ψ, is of the special importance. In this case, the technique of the perturbed Lyapunov function makes it possible to investigate the stability on a finite interval and to estimate the length of the time interval on which the integral curves will remain in the ϵ – neighbourhood.

While the negativeness of the mean of v enables us to prove theorems on stability, the positiveness of the mean of derivative of the perturbed Lyapunov function v makes it possible to generalize the Lyapunov and Chetayev theorems on instabilities to this subtle case and presents a convenient technique for detecting resonances in nonlinear systems.

Many complex situations are associated with the mean (1.7). Under certain additional conditions, we have attraction of a solution to the equilibrium position and the asymptotic stability. It becomes easier to check whether the mean is negative if the unperturbed system splits into subsystems in the first approximation and additional conditions concerning the mean values of the subsystems are satisfied. Below, we consider interesting questions concerning the non – uniform character of convergence of the mean to zero when the system approach the equilibrium position and the possibility of averaging over time which enters the equations explicitly. Systems with weak quasi – periodic autonomy are also considered. The ideas presented above are applied in the studies of stability of integro – differential systems, where we propose and examine the new averaging schemes and equations with kernels of the difference type which are typical for applied problems, in particular, the problems of visco – elasticity.

The main theorems concerning the generalized Lyapunov method will be formulated below in a special form, as to make convenient to check numerically if their assumptions are satisfied.

The generalized Lyapunov technique is used to analyze systems with nonlinearities which are quasi – periodic in time, under assumption that the constant matrix of the linear approximation has imaginary eigenvalues only. We consider resonances up to the order four; to do that, we construct the perturbed Lyapunov function.

In conclusion, we prove a general theorem which does not make use of the Lyapunov function of the unperturbed system; we use a positive – definite function and the assumption that the equilibrium position of the unperturbed system is stable, instead. This theorem, allows us to use the technique of Lyapunov vector functions to study stability of the system.

2.2 Lyapunov Functions Positive – Definite in Subset of Variables

In the present chapter, we study stability with respect to subset of all variables. Such approach enlarges a circle of applied problems, to which the generalized Lyapunov method can be applied. Some problems of celestial mechanics, such as the classical three – body problem or the problem of motion of a satellite with respect to the mass centre can be examined. In gyroscopic systems some of the phases are difficult to observe. Therefore, in dealing with such problems, the study of stability usually takes into consideration only part of all variables. In problems of gyroscope stability we find the conditions of stability containing only the observed (measurable) parameters.

We describe the equilibrium position of the unperturbed system by means of the Lyapunov function, which is positive – definite in a subset of variables only. We present the definition of such functions, following [94], [97].

Consider a function $v_0(t, x, y)$, where x and y are vectors, $\dim x = n$, $\dim y = m$ in the domain $P = P\{t \geq 0, \| x \| < h, y \in D\}$, and the function v_0 has continuous derivative in P and satisfies the condition $v_0(t, 0, y) = 0$.

Definition 2.1 *The function $v_0(t, x, y)$ admits infinitely small upper bound in the variable x if for each $\delta > 0$ there exists $\gamma > 0$ such that for all t, x, y*

$$\| x \| < \gamma, \quad t > 0, \quad y \in D$$

the inequality $|v_0(t, x, y)| < \delta$ is satisfied.

Definition 2.2 *The function $v_0(t, x, y)$ is positive – definite in the variable x in the domain P if it satisfies the inequality*

$$w(x) \leq v_0(t, x, y) \tag{2.1}$$

where $w(x)$ is a positive – definite function which is independent of t and y.

To prove Lyapunov theorems, one must consider manifolds defined by the equation $v = $ const. If the Lyapunov function depends on x

only, then the equation $v(x) = c$ defines a surface. In stability theory, the manifolds $v = $ const are traditionally called equipotential surfaces also in the case of the non – stationary motion, when the Lyapunov function depends on t as well, cf. [66], [18], [75].

In studies of stability with respect to the subset of variables, the manifold $v(t, x, y) = $ const in the space of variables x becomes much more complicated. But, as in the case of $v = v(t, x)$, one can speak about a system of equipotential surfaces depending on the parameters t and y, or about the moving equipotential surfaces. Here, we follow tradition and in the following we will call the manifold in the space $v(t, x, y) = $ const the equipotential surface.

Similarly to the case of unstabilized motions with $v = v(t, x)$, one can give a geometric interpretation of the surface $v_0(t, x, y) = c$. In particular, it is easy to establish that for a positive – definite function which possesses an infinitely small upper bound, the equipotential surface $v_0(t, x, y) = c$ $(c > 0)$ lies, in variable x, outside some neighbourhood of the origin.

We note also that, since the function is positive – definite, the moving equipotential surfaces $v_0(t, x, y) = c$ lie inside an static equipotential surface.

2.3 Equipotential Surfaces of Perturbed Lyapunov Function. Proximity of Solutions of Complete and Unperturbed Systems

1. Here we consider the perturbed Lyapunov function $v = v_0(t, x, y) + u(t, x, y, \mu, \epsilon)$, where $v_0(t, x, y)$ is the Lyapunov function of the system (1.3) being positive – definite in the domain P, which possesses an infinitely small upper bound in the variable x, and $u(t, x, y, \mu, \epsilon)$ is a perturbation of the Lyapunov function. Equipotential surfaces (manifolds) of such functions are introduced. The lemma on properties of these surfaces is formulated.

Before formulating this lemma, we carry out some preliminary constructions. We fix $\bar{\epsilon} > 0$ ($\bar{\epsilon} < H$) and chose the number $w_0 > 0$ such that the surface $w(x) = w_0$ lies in the $\bar{\epsilon}$ – neighbourhood of the point x. Let the number $\sigma > 0$ satisfy the condition $2\sigma < w_0$. Moreover, since v_0 is positive – definite with respect to x, the surfaces

$$v_0(t, x, y) = w_0 \tag{3.1}$$

$$v_0(t, x, y) = w_0 - 2\sigma \tag{3.2}$$

lie in the surface $w(x) = w_0$.

Since the function v_0 possesses an infinitely small upper bound in the variable x and there exists $\eta > 0$ such that η – neighbourhood is entirely contained in the surfaces (3.1) and (3.2) (see Sect. 2).

Lemma 2.1 *Let the perturbation $u(t, x, y, \mu, \epsilon)$ of the Lyapunov function be a continuous function with the modulus bounded by $\sigma/2$ in the annular region $\eta \leq \| x \| \leq \bar{\epsilon}$, $t > 0$, $y \in D$. Then the points x belonging to the surface*

$$v_0(t, x, y) + u(t, x, y, \mu, \epsilon) = w_0 - \sigma \tag{3.3}$$

satisfy the condition $\eta \leq \| x \| \leq \bar{\epsilon}$ and this surface is closed.

Proof. For x belonging to the annulus $\eta \leq \| x \| \leq \bar{\epsilon}$, the function $v_0(t, x, y)$ takes the values $w_0 - 2\sigma$ on the surface (3.2) and w_0 on the surface (3.1), respectively.

On the surface (3.2), the continuous function $v_0 + u = v$ takes the value v^*, which is bounded, to wit

$$w_0 - 2\sigma - \frac{1}{2}\sigma \leq v^* \leq w_0 - 2\sigma + \frac{1}{2}\sigma. \tag{3.4}$$

On the surface (3.1), the function v takes the value v^{**} which belongs to the interval

$$w_0 - \frac{1}{2}\sigma \leq v^{**} \leq w_0 + \frac{1}{2}\sigma. \tag{3.5}$$

Since v is continuous, it takes also the intermediate value $w_0 - \sigma$ in the annular region. Thus, the surface (3.3) lies also in the annular region $\eta \leq \| x \| \leq \bar{\epsilon}$ in the variable x.

Let us prove now that Eq. (3.3) defines a closed surface lying in the annulus $\eta \leq \| x \| \leq \bar{\epsilon}$ in the variable x. To this end, we draw an arbitrary continuous curve C from the surface $\| x \| = \eta$ to the surface $w(x) = w_0$. On this curve there is a point C^*, belonging to the surface (3.2), where v takes the value v^* (3.4), and the point C^{**}, where v takes the value v^{**} (3.5). Since C is continuous, there exists a point on the segment of the curve between C^* and C^{**}, where v takes the intermediate value, because $v^* < w_0 - \sigma < v^{**}$. But C is an arbitrary continuous curve, and therefore, Eq. (3.3) defines a closed surface lying in the annulus $\eta \leq \| x \| \leq \bar{\epsilon}$ and inside the surface $w(x) = w_0$ in the variable x.

2. Let us turn now to the question of proximity of solutions of the complete and unperturbed systems.

Lemma 2.2 *Let the RHS of the systems (1.1) and (1.3) satisfy the conditions ensuring existence and uniqueness of solutions in the domain P. Let $f(t, z)$ satisfy the Lipschitz condition with the constant N with respect to z. Moreover, let a summable function $M(t)$ and a constant M_0 exist such that in the domain P, on any finite segment $[t_1, t_2]$*

$$\| R(t, z, \mu) \| < M(t); \quad \int_{t_1}^{t_2} M(t)\, dt \leq M_0(t_2 - t_1). \quad (3.6)$$

Then, for the solution $z = z(t)$ of the system (1.1) and the solution $z = \bar{z}$ of the system (1.3), both starting from the point $z(t_0) = \bar{z}(t_0) = z_0$, the inequality

$$\| z(t) - \bar{z}(t) \| < \mu M_0 l e^{NL} \quad (3.7)$$

holds for t belonging to the interval $[t_0, t_0 + l]$.

Proof. The system (1.1) is equivalent to the following system of integral equations

$$z(t) = z_0 + \int_{t_0}^t f(t, z(t))\, dt + \mu \int_{t_0}^t R(t, z(t), \mu)\, dt. \qquad (3.8)$$

Let us write down the system of integral equations which is equivalent to the unperturbed system (1.3) and has the same initial data

$$\bar{z}(t) = z_0 + \int_{t_0}^t f(t, \bar{z}(t))\, dt. \qquad (3.9)$$

Subtracting these two equations, we obtain

$$z(t) - \bar{z}(t) = \int_{t_0}^t [f(t, z(t)) - f(t, \bar{z}(t))]\, dt + \mu \int_{t_0}^t [R(t, z(t), \mu)\, dt.$$

Using the Lipschitz inequality for f and the fact that R is bounded by a summable function, we find

$$\| z - \bar{z} \| \leq N \int_{t_0}^t \| z - \bar{z} \|\, dt + \mu M_0 l \qquad (3.10)$$

which holds in the domain P and for $t_0 \leq t \leq t_0 + l$.

Combining (3.10) and the Gronwall lemma [99], we obtain

$$\| z - \bar{z} \| \leq \mu M_0 l e^{Nl}$$

QED.

2.4 Perturbed Lyapunov Function in Annular Region. Theorem on Stability

1. To prove the theorem below, we will assume that there exists a perturbed Lyapunov function

$$v = v_0(t, x, y) + u(t, x, y, \mu, \epsilon)$$

where the perturbation u is given not in the entire domain P, but only in the annular region $0 < \eta \leq \| x \| \leq \epsilon$, $t > 0$, $y \in D$, centered at the point $x = 0$. In the η – neighbourhood of the equilibrium position, the perturbation u may be not defined at all.

First, we carry out some preliminary constructions needed to formulate this theorem.

Let us consider a positive – definite function $v_0(t, x, y)$ that possesses an infinitely small upper bound in the variable x. According to Definition 2.2, there exists a positive – definite function $w(x)$ such that

$$v_0(t, x, y) \geq w(x).$$

Next, we fix the numbers ϵ and $\bar{\epsilon}$ satisfying the condition $\bar{\epsilon} < \epsilon < H$. We choose a number w_0 such that the static surface $w(x) = w_0$ lies entirely in the $\bar{\epsilon}$ – neighbourhood of the point $x = 0$ and consider the moving surfaces

$$v_0(t, x, y) = w_0, \quad v_0(t, x, y) = w_0 - 2\sigma_0$$

where σ_0 satisfies the inequality $2\sigma_0 < w_0$. By virtue of the inequality $v_0(t, x, y) \geq w(x)$, both these surfaces lie inside the static surface $w(x) = w_0$. The Lyapunov function v_0 possesses an infinitely small upper bound, and therefore one can find $\gamma > 0$ such that for all $t > 0$, $y \in D$, the γ – neighbourhood lies inside the moving surfaces introduced above.

To simplify application of the theorem to resonance problems, we introduce restrictions, related to the derivative of u, on the sign of the integral $J(t_0, x_0, y_0, T, \mu, \epsilon)$. This integral was introduced in Eq. (1.7) and should be calculated by integration along the solution $x = \bar{x}(t)$, $y = \bar{y}(t)$ of the unperturbed system (1.3), since the limit leading to the mean (1.8) may not exist in such problems.

The results of this section were first published in [53], [43].

Theorem 2.1 *Suppose that:*

a) There exists a Lyapunov function of the system (1.3) $v_0(t, x, y)$ *which is positive – definite with respect to the x variable and possessing an infinitely small upper bound in this variable.*

b) There exists a total derivative of v_0, constructed according to Eq. (1.3) which is non – positive in the domain P.

Moreover, assume that for any $\epsilon > 0$ ($\epsilon < \epsilon_0 < H$) one can choose w_0, σ_0, and $\gamma(\sigma_0 < \frac{1}{2}w_0, \gamma(w_0, \sigma_0))$ related, as mentioned in Sect. 2.1 through v_0, such that for all $\rho \geq \gamma$ ($\rho < \epsilon$) the functions $u(t, x, y, \mu, \epsilon)$ and $\phi(t, x, y, \mu, \epsilon)$ are defined in the annulus $\rho <\| x \|< \epsilon$, $t > 0$, $y \in D$, and

c) There exist $\delta > 0$ and $l > 0$ such that if for $t_0 \leq t \leq t_0 + T$ $\rho <\| \bar{x} \|< \epsilon$, then

$$\int_{t_0}^{t_0+T} \phi(t, \bar{x}(t), \bar{y}(t), \mu, \epsilon) \, dt < -\delta T$$

uniformly with respect to $t_0 > 0$ and $y_0 \in D$ ($T > 1$).

d) There exist two summable functions $F(t)$, $M(t)$ and constants F_0, M_0, and a non – decreasing function $\chi_1(\alpha)$, $\lim_{\alpha \to 0} \chi_1(\alpha) = 0$ such that for $\rho <\| x \|\leq \epsilon$, $t > 0$, $y \in D$

$$|\phi(t, z') - \phi(t, z'')| \leq \chi_1(|z' - z''|) F(t)$$
$$\int_{t_1}^{t_2} F(t) \, dt \leq F_0(t_2 - t_1) \qquad \int_{t_1}^{t_2} M(t) \, dt \leq M_0(t_2 - t_1)$$
$$\| R(t, z) \|\leq M(t)$$

on any finite interval $[t_1, t_2]$.

e) There exists a non – decreasing function $\chi_2(\mu) > 0$, $\lim_{\mu \to 0} \chi_2(\mu) = 0$ such that $|u(t, z, \mu, \epsilon)| < \chi_2(\mu)$ for $\rho <\| x \|< \epsilon$.

Under these conditions, there exist $\eta(\epsilon)$ and $\mu_0(\epsilon)$ such that for any solution $x = x(t)$, $y = y(t)$ of the system (1.1) with the initial condition $x(0) = x_0$, $y(0) = y_0$ satisfying $\| x_0 \|< \eta$, the inequality $\| x(t) \|< \epsilon$ holds for $\mu < \mu_0$ and all $t > 0$.

Before start proving the theorem, it is worth noticing that the case $u = 0$ (i.e. $v = v_0$) is more simple. Hence, we would like to recommend the reader to read first the formulation of Theorem 2.5, Sec. 2.7 which is a particular case of Theorem 2.1.

Proof of Theorem 2.1. We fix $\bar{\epsilon} > 0$, $\bar{\epsilon} < \epsilon$ and consider the surfaces (3.1) and (3.2). If one chooses a positive σ which satisfies $\sigma < \sigma_0$, then, according to condition (e) of the theorem, there exists $\eta \geq \gamma$ such that the η-neighbourhood of the point $x = 0$ lies inside the surfaces

$$v_0(t, x, y) = w_0, \quad v_0(t, x, y) = w_0 - 2\sigma. \tag{4.1}$$

Moreover, in the annular region $\eta < \| x \| < \epsilon$ assumptions (c), (d), and (e) hold.

Consider the surface

$$v = v_0(t, x, y) + u(t, x, y, \mu, \epsilon) = w_0 - \sigma. \tag{4.2}$$

Condition (e) enables us to make μ_1 small enough to ensure that for all μ, $0 < \mu < \mu_1$, the inequality $|u| < \frac{1}{2}\sigma$ holds. Then, according to Lemma 2.1, the surface (4.2) is closed and lies in the annular region $\eta \leq \| x \| < \bar{\epsilon}$, $t > 0$, $y \in D$.

Let us construct the integral curve of the system (1.1): $x = x(t)$, $y = y(t)$ from a point lying in the η-neighbourhood of the equilibrium position. This curve starts from the region $\| x \| < \eta$ and at some time $t = t_0$ intersects the closed surface (4.2). Let us trace change of the perturbed Lyapunov function v along the solution $x = x(t)$, $y = y(t)$. To this end, we differentiate the function v along this solution $z = z(t)$ and, using the notation introduced in (1.6), we get

$$\frac{dv}{dt} = \frac{\partial v_0}{\partial t} + \frac{\partial v_0}{\partial z} f + \theta(\mu)\phi(t, z, \mu, \epsilon). \tag{4.3}$$

Let us integrate (4.2) along the solution $z = z(t)$, taking condition (b) of the theorem

$$\frac{\partial v_0}{\partial t} + \frac{\partial v_0}{\partial z} f \leq 0$$

into account. As a result, we get the inequality

$$v(t, z(t)) \leq v(t_0, z_0) + \theta(\mu) \int_{t_0}^{t} \phi(t, z(t), \mu, \epsilon) \, dt. \tag{4.4}$$

Now, we construct the integral curve $z = \bar{z}(t)$ of the unperturbed system (1.3) which starts from the same point t_0, z_0, and then we add and subtract the integral calculated along this curve from the RHS of the inequality (4.4), to obtain

$$\int_{t_0}^{t} \phi(t, z(t), \mu, \epsilon)\, dt = \int_{t_0}^{t} \phi(t, \bar{z}(t), \mu, \epsilon)\, dt +$$

$$+ \int_{t_0}^{t} (\phi(t, z(t), \mu, \epsilon) - \phi(t, \bar{z}(t), \mu, \epsilon))\, dt. \tag{4.5}$$

Below, for the sake of brevity, we will omit the arguments μ and ϵ of the function ϕ.

According to condition (c) of the theorem, there exist positive numbers δ and l such that if $t > t_0 + l$ and if the condition $\eta \le \bar{x}(\tau) \parallel < \epsilon$ holds for $\tau \in [t_0, t]$, then

$$\int_{t_0}^{t} \phi(t, \bar{z}(t))\, dt \le -\delta(t - t_0). \tag{4.6}$$

uniformly with respect to t_0.

Let us examine the evolution of v for $t_0 < t < t_0 + l$, where $l > 0$ is determined by condition (c) of the theorem.

Assumptions (a) and (b) of the theorem determine stability of the point $x = 0$ for the unperturbed system (1.3). Therefore, the curve $z = z(t)$ will remain, in the variable x, in the $\bar{\epsilon}$-neighbourhood of the point $x = 0$ for all $t > t_0$.

Lemma 2.2 gives us the estimate of deviation of solutions of the perturbed and unperturbed systems (1.1) and (1.3), the solutions which start from the same point. If, by choosing the small parameter μ, one makes this deviation smaller than $\epsilon - \bar{\epsilon}$ on the time interval of length l, then the solution $z = z(t)$ of the perturbed system will not leave the ϵ-neighbourhood of the point $x = 0$ on this time interval. Let

$$\mu_2 = (\epsilon - \bar{\epsilon})[M_0 l e^{Nl}]^{-1}. \tag{4.7}$$

Then, according to Lemma 2.2, for every μ, $0 < \mu < \mu_2$ and $t \in [t_0, t_0+l]$

$$\| x(t) - \bar{x}(t) \| \le \epsilon - \bar{\epsilon}. \tag{4.8}$$

Two cases are possible here: (1) the curve remains outside the η-neighbourhood in variables x for $t_0 \le t \le t_0 + l$; or (2) there exists a point $t_1 \in [t_0, t_0 + l]$ such that $\| x(t_1) \| \le \eta$.

First, we consider case (2), when $v_0(t_1, \bar{z}(t_1)) < w_0 - 2\sigma$; for v we find $v(t_1, z(t_1)) = v_0(t_1, z(t_1)) + u = v_0(t_1, \bar{z}(t_1)) + u + v_0(t_1, z(t_1)) - v_0(t_1, \bar{z}(t_1))$. Here, the second term satisfies $|u| < \frac{1}{2}\sigma$. To estimate the difference of the Lyapunov function at the points $z(t_1)$ and $\bar{z}(t_1)$ we use continuity of the function v_0 and the bound for deviation of the solutions $\bar{z}$ and z emanating from the same point on the time interval $[t_0, t_0 + l]$ given in Lemma 2.2. We choose $\bar{\mu}_2$ small enough, as to ensure the inequality

$$|v_0(t_1, \bar{z}(t_1)) - v_0(t_1, z(t_1))| < \frac{1}{2}\sigma$$

for $t_1 \in [t_0, t_0 + l]$. Here, $v_0(t_1, z(t_1)) < w_0 - \sigma$ and the curve remains inside the surface (4.2) for $t = t_1$.

Let us turn to case (1). According to condition (d) of the theorem, the inequality

$$|\phi(t, z(t)) - \phi(t, \bar{z}(t))| \le \chi(|z(t) - \bar{z}(t)|)F(t) \tag{4.9}$$

holds.

Estimating the difference $z - \bar{z}$ with the help of Lemma 2.2, we obtain

$$(t - t_0)^{-1} \int_{t_0}^{t} |\phi(t, z(t)) - \phi(t, \bar{z}(t))|\, dt \le$$

$$\tag{4.10}$$

$$\le \chi_1(\mu M_0 l e^{Nl})F_0$$

for the interval $[t_0, t_0 + l]$. According to condition (c) of the theorem, there exists μ_3 small enough, as to ensure the inequality

$$\chi_1(\mu_3 M_0 l e^{Nl})F_0 \le \frac{1}{2}\delta. \tag{4.11}$$

In addition, the second integral on the RHS of (4.5) satisfies the inequality

$$\left| \int_{t_0}^{t} [\phi(t, z(t)) - \phi(t, \bar{z}(t))] \, dt \right| < \frac{1}{2} \delta (t - t_0) \tag{4.12}$$

for $t \in [t_0, t_0 + l]$.

If one chooses $\mu_0 = \min\{\mu_1, \mu_2, \bar{\mu}_2, \mu_3\}$, then for every μ, $0 < \mu < \mu_0$ and $t_0 \leq t \leq t_0 + l$ one ensures the inequalities $\| x(t) \| < \epsilon$, $|u| < \frac{1}{2}\sigma$, and (4.12).

Combining the inequality (4.12), condition (c) of the theorem, and Eqs. (4.5), (4.6), we can estimate the integral $\int_{t_0}^{t} \phi(t, z(t)) \, dt$ in the inequality (4.4) for $t_0 + l \leq t$:

$$\int_{t_0}^{t} \phi(t, z(t)) \, dt \leq (t - t_0)(-\delta + \frac{1}{2}\delta). \tag{4.13}$$

Thus, the solution $z = z(t)$ remains, for $t \in [t_0, t_0 + l]$ in the ϵ-neighbourhood in the variables x; along this solution, at least starting from some point of the interval $[t_0, t_0 + l]$ on, the integral $\int_{t_0}^{t} \phi(t, z(t)) \, dt$ becomes negative and the function decreases. This means that the solution returns to the surface (4.2), as in the case considered before. All the estimates are uniform with respect to t_0, z_0, therefore, the solution may leave the surface (4.2) an unlimited number of times and then return: for every $t > 0$ it remains in the region $\| x \| < \epsilon$. QED.

If the perturbed Lyapunov function $v = v_0 + u$ is the integral of the system (1.1), i.e. if $\phi = 0$, then we have

Theorem 2.2 *Assuming that conditions (a), (c), and (e) of Theorem 2.1 are satisfied and $\phi = 0$ for $\rho < \| x \| < \epsilon$, $t > 0$, $y \in D$, the assertion of Theorem 2.1 is true.*

Proof. We use the system (4.1) – (4.2) of equipotential surfaces, constructed in proof of Theorem 2.1.

Condition (e) of Theorem 2.1 makes it possible to choose $\mu_0(\epsilon)$ such that the inequality $|u| < \frac{1}{2}\sigma$ holds in the annular region $\eta < \| x \| < \epsilon$.

According to condition (b) of Theorem 2.1 and condition $\phi = 0$ of Theorem 2.2, the perturbed Lyapunov function $v(t, x, y, \mu, \epsilon)$ does not increase along the integral curve emanating from the η-neighbourhood of the point $x = 0$. Therefore, the integral curve $x = x(t)$, $y = y(t)$, having the initial data $x(0) = x_0$, $y(0) = y_0$ and satisfying the inequality $\| x \| < \eta$, will never intersect the surface (4.2) and hence, it will never leave the ϵ-neighbourhood of the point $x = 0$. QED.

2.5 Theorem on Attraction of Solutions to Equilibrium Point

If the assumptions of Theorem 2.1 hold, one can use more restrictive estimates to show that the solution x does not only remain within ϵ-neighbourhood of the equilibrium position, but is also contracted into a smaller λ-neighbourhood. To prove this statement, one has to study evolution of the perturbed Lyapunov function v along the integral curves of the system (1.1) lying outside some γ-neighbourhood of the equilibrium position. Conditions of Theorem 2.1 enable us to construct a system of moving surfaces in the λ-neighbourhood ($\lambda \leq \epsilon$) and choose a $\gamma(\lambda)$-neighbourhood of the point $x = 0$ outside which the conditions regarding the sign of the derivative of v are satisfied. To prove asymptotic stability, one has to assume that the functions $R(t, z)$ are equal to zero at the point $x = 0$.

Theorem 2.3 *Let the assumptions of Theorem 2.1 be satisfied. Then for any solution $x(t)$, $\| x_0 \| < \eta$ and for any λ ($\lambda < \epsilon$) there exist $\mu_0(\lambda)$ and $T(\lambda)$ such that $\| x(t) \| \leq \lambda$ for $t > T(\lambda)$ and $0 < \mu < \mu_0$.*

Proof. In the λ-neighbourhood of the point $x = 0$ we construct a system of surfaces containing the equilibrium position. To this end, we choose w_λ such that the surface $w(x) = w_\lambda$ lies in the λ-neighbourhood of the equilibrium position $x = 0$. Let σ_λ be such that $0 < 2\sigma_\lambda < w_\lambda$.

Consider the equipotential surfaces

$$v_0(t, x, y) = w_\lambda - 2\sigma_\lambda, \quad v_0(t, x, y) = w_\lambda. \tag{5.1}$$

Since the function v_0 is positive – definite and admits an infinitely small upper bound, there exists $\bar{\eta}(\lambda)$ such that the points satisfying the condition $\| x \| < \bar{\eta}(\lambda)$ lie inside the equipotential surface (5.1). We choose μ_1 such that $|u(t, x, y, \mu, \epsilon)| < \frac{1}{2}\sigma_\lambda$. Then the surface

$$v_0(t, x, y) + u(t, x, y, \mu, \epsilon) = w_\lambda - \sigma_\lambda \tag{5.2}$$

lies inside the annular region

$$\bar{\eta}(\lambda) < \| x \| < \lambda. \tag{5.3}$$

Let us examine the evolution of the generalized (perturbed) Lyapunov function $v = v_0 + u$ along the solution $z = z(t)$ of the system (1.1). If the initial point satisfies in the variables x the condition $\| x \| < \bar{\eta}$, then we get the assertion of the theorem by applying Theorem 2.1. Consider the initial data x_0 such that $\bar{\eta} < \| x \|$.

Taking into account the fact that $\dfrac{\partial v_0}{\partial t} + \nabla v_0 \, f \leq 0$, we get

$$v(t, z(t)) \leq v(0, z(0)) + \theta(\mu) \int_0^t \phi(t, z, \mu, \epsilon) \, dt. \tag{5.4}$$

We represent the integral from Eq. (5.4) as a sum of integrals over the segments $[t_k, t_{k+1}]$ and then add and subtract the integrals along the solutions $\bar{z}_k(t)$ of the system (1.3) with the initial data $z(t_k) = \bar{z}(t_k) = z_{k_0}$, to wit

$$\int_0^t \phi(t, z(t)) \, dt =$$

$$\tag{5.5}$$

$$= \sum_{k=0}^S \left\{ \int_{t_k}^{t_{k+1}} \phi(t, \bar{z}_k(t)) \, dt + \int_{t_k}^{t_{k+1}} [\phi(t, z(t)) - \phi(t, \bar{z}_k(t))] \, dt \right\}.$$

Let us assume that the curve $z = z(t)$ does not enter the $\bar{\eta}$-neighbourhood of the point $x = 0$ for any positive t.

We choose l sufficiently large that for $l \leq t_{k+1} - t_k \leq 2l$ we have

$$\int_{t_k}^{t_{k+1}} \phi(t, \bar{z}_k(t)) \, dt < -\bar{\delta}(\lambda) \Delta t_k. \tag{5.6}$$

According to assumption (c) of Theorem 2.1 such $\bar{\delta}(\lambda)$ does exist. The second integral on the RHS of (5.5) is estimated with the help of condition (d) of Theorem 2.1:

$$\int_{t_k}^{t_{k+1}} [\phi(t, z(t)) - \phi(t, \bar{z}_k(t))] \, dt \leq$$

$$\tag{5.7}$$

$$\leq \chi_1(\mu M_0 2l \exp(2Nl)) F_0 \Delta t_k.$$

We choose μ_1 sufficiently small to guarantee the inequality

$$\chi_1(\mu M_0 2l \exp(2Nl)) F_0 \leq \frac{1}{2} \bar{\delta}. \tag{5.8}$$

Combining inequalities (5.4), (5.6), and (5.8), we arrive at

$$v(t, z(t)) - v(0, z(0)) \leq -\theta(\mu) \sum_{k=1}^{S} \frac{\bar{\delta}}{2} \Delta t_k. \tag{5.9}$$

When S tends to ∞, then, starting from some $t = t^*$, the Lyapunov function v_0 becomes negative, which contradicts the condition that $v_0(t, z)$ is positive – definite. Hence, the solution should enter the $\bar{\eta}$-neighbourhood in the variables x and, by virtue of Theorem 2.1, remain in the λ-neighbourhood. QED.

The theorem proved above asserts that if μ is sufficiently small, then the solution is contracted into an arbitrarily small λ-neighbourhood.

This result was first published in [43]; it was proved in [61] for the functions $v = v_0 + u$ defined not in the annular region, but in a certain neighbourhood of the equilibrium position.

2.6 Investigation of Stability on Finite Interval

The negativeness of either the integral $J(t_0, z_0, T, \mu, \epsilon)$, or the mean $\psi(t_0, z_0, \mu, \epsilon)$ of the derivative of perturbed Lyapunov function v, or the fact that the derivative is zero, enabled us to prove Theorems 2.1 and 2.2 on stability. Otherwise, when either the derivative $\dot{v}$ is non-zero, or the signs of ψ and/or $J(t_0, z_0, T, \mu, \epsilon)$ are unknown, for the perturbed system (1.1) the point $x = 0$ may be unstable. In such cases, however, using the techniques of perturbed Lyapunov function, one can estimate the interval of time during which the solution, initially close to the stationary point, remains in its fixed vicinity. Here, the dependence of the length of the segment on both the size of neighbourhood of equilibrium position and the size of perturbations is important. Conditions of Theorem 2.4, which we are going to prove, are substantially weaker than the conditions of Theorems 2.1 and 2.2, and this makes Theorem 2.4 useful for practical applications. We note that this assertion, as well as the theorems concerning the averaging method provides the estimate of proximity of solutions (here with respect to the equilibrium position) on the time interval measured in terms of inverse powers of μ. This is a typical situation in problems admitting averaging under additional condition that the unperturbed system possesses an equilibrium position.

Theorem 2.4 *Let the following conditions be satisfied:*

a) There exists a Lyapunov function of the system (1.3) $v_0(t, x, y)$ *which is positive – definite with respect to the x variable and possesses an infinitely small upper bound in this variable.*

b) There exists a total derivative of v_0, constructed according to Eq. (1.3) *which is non – positive in the region* (1.2).

Moreover, assume that for any $\epsilon > 0$ and $\rho > 0$ ($\rho < \epsilon$) the functions $u(t, x, y, \mu, \epsilon)$ and $\phi(t, x, y, \mu, \epsilon)$ are defined in the annular region $\rho < \| x \| < \epsilon$, $t > 0$, $y \in D$, and

c) There exists a non – decreasing function $\chi_1(\alpha)$, $\lim\limits_{\alpha \to 0} \chi_1(\alpha) = 0$ such that for $\rho < \| x \| < \epsilon$, $t > 0$, $y \in D$

$$|u(t, x, y, \mu, \epsilon)| < \chi_1(\alpha).$$

d) There exists a constant ϕ_0 such that $|\phi(t,x,y,\mu,\epsilon)| < \phi_0$ for $\rho <\| x \|< \epsilon$, $t > 0$, $y \in D$.

Then there exist $\sigma(\epsilon)$, $T(\epsilon,\mu) = \sigma(\epsilon)[2\phi_0\theta(\mu)]^{-1}$, $\eta(\epsilon)$, $\mu_0(\epsilon)$ such that all solutions which at the initial moment satisfy the inequality $\| x_0 \|< \eta$ in the variable x satisfy the inequality $\| x(t) \|< \epsilon$ for all $0 < t < T(\epsilon,\mu)$ and $\mu < \mu_0(\epsilon)$ as well.

Proof. Let $\epsilon > 0$ be fixed. According to condition (a) of Theorem 2.4, there exists a positive – definite function $w(x)$ such that

$$v_0(t,x,y) \geq w(x). \tag{6.1}$$

We choose w_0 such that the surface $w(x) = w_0$ lies entirely in the ϵ-neighbourhood of the point $x = 0$. Consider the moving surfaces

$$v_0(t,x,y) = w_0, \quad v_0(t,x,y) = w_0 - 2\sigma, \tag{6.2}$$

where σ is some number satisfying the condition $0 < 2\sigma < w_0$. By virtue of the inequality (6.1), the moving surfaces (6.2) lie inside the static surface $w(x) = w_0$.

Since v_0 admits an infinitely small upper bound, there exists an η-neighbourhood of the point $x = 0$ which lies inside the surfaces (6.2) for all $t > 0$, $y \in D$.

Now, we consider the surface

$$v = v_0 + u = w_0 - \sigma \tag{6.3}$$

and choose μ_0 small enough as to guarantee that for every $\mu < \mu_0$ the inequality

$$|u(t,x,y,\mu,\epsilon)| < \frac{\sigma}{4} \tag{6.4}$$

holds; this is possible due to condition (c).

Let us construct the integral curve $x = x(t)$, $y = y(t)$ starting from the η-neighbourhood of the stationary point and assume that at the moment $t = t_0$ it intersects the surface (6.3):

$$v|_{t=t_0} = w_0 - \sigma. \tag{6.5}$$

According to Lemma 2.1, the surface (6.5) is closed and lies outside the η-neighbourhood of the point $x = 0$; therefore, in the variables x, this curve can leave the ϵ-neighbourhood of the point $x = 0$ only after intersecting the surface (6.5).

Let us differentiate the function v and apply Eq. (1.1). Using notation (1.5) and (1.6), we obtain

$$\frac{dv}{dt} = \frac{\partial v_0}{\partial t} + \frac{\partial v_0}{\partial z} f + \phi(t, z, \mu, \epsilon)\theta(\mu). \tag{6.6}$$

Integrating (6.6) from the moment $t = t_0$ and taking (6.5) and condition (b) of Theorem 2.4 into account, we obtain

$$v \leq w_0 - \sigma + U, \tag{6.7}$$

where

$$U(t) = \theta(\mu) \int_{t_0}^{t} \phi(t, z(t), \mu, \epsilon)\, dt, \tag{6.8}$$

i.e.

$$v_0 + u \leq w_0 - \sigma + U(t), \tag{6.9}$$

or $v_0 \leq w_0 - \sigma - u + U(t)$.

Combining (6.8) and condition (d) of the theorem, we make sure that U does not exceed $\frac{1}{2}\sigma$ on the interval $[t_0, t_0 + T]$, with $T = \sigma[\phi_0\theta(\mu)]^{-1}$, and therefore, v_0 does not exceed w_0.

Since the moving surface $v_0 = w_0$ lies inside the static surface $w(x) = w_0$, for $t \in [t_0, t_0 + T]$ the integral curve will not leave the ϵ-neighbourhood of the point $x = 0$. The theorem is proved.

2.7 Investigation of Stability in Higher Approximations

1. In the cases of practical importance (see, e.g. Sect. 1) a special form of Theorem 2.1, when the perturbation u of the Lyapunov function is

equal to zero identically and there exists the mean (1.8) is particularly useful. Here, the Lyapunov function is positive definite, but its derivative may change the sign if the sign of the mean $\psi(t_0, z_0, \mu, \epsilon)$ is negative. Since this particular case is of special importance, we formulate the corresponding theorem (cf. [53]).

Theorem 2.5 *Suppose that:*

a) There exists a Lyapunov function of the system (1.3) $v_0(t, x, y)$ *which is positive − definite with respect to the x variable and possessing an infinitely small upper bound in this variable.*

b) There exists a total derivative of v_0, constructed according to Eq. (1.3) *which is non − positive in the domain P.*

c) The mean

$$\psi_0(t_0, x_0, y_0) =$$
$$= \lim_{T \to \infty} \frac{1}{T} \int_{t_0}^{t_0+T} \frac{\partial v_0(t, \bar{z}(t))}{\partial z} R(t, \bar{z}(t), 0) \, dt$$

exists uniformly with respect to t_0, x_0, y_0 from the domain P.

Moreover, for every $\gamma > 0$ ($\gamma < H$) there exists $\delta > 0$ such that if $\| x_0 \| > \gamma$, *then* $\psi_0(t_0, x_0, y_0) < -\delta$ *for* $t_0 \geq 0$, $y_0 \in D$.

d) There exist two summable functions $F(t)$, $M(t)$ and constants F_0, M_0, and a non − decreasing function $\chi_1(\alpha)$, $\lim_{\alpha \to 0} \chi_1(\alpha) = 0$ such that in the domain (1.2) *we have*

$$|\phi(t, z') - \phi(t, z'')| \leq \chi_1(|z' - z''|)F(t),$$
$$\int_{t_1}^{t_2} F(t) \, dt \leq F_0(t_2 - t_1), \qquad \int_{t_1}^{t_2} M(t) \, dt \leq M_0(t_2 - t_1),$$
$$\| R(t, z) \| \leq M(t).$$

on any finite interval $[t_1, t_2]$.

Then, for any $\epsilon > 0$ there exist $\eta > 0$ and $\mu_0(\epsilon)$ such that for any solution of the system (1.1) *with the initial condition $x(0) = x_0$, $y(0) = y_0$ satisfying $\| x_0 \| < \eta$, the inequality $\| x(t) \| < \epsilon$ holds for $\mu < \mu_0$ and all $t > 0$.*

2. Let us now construct a perturbation $u(t, z, \mu)$ of the Lyapunov function; the perturbation will be constructed as a polynomial or series in powers of a small parameter. For the sake of simplicity, we perform the construction not in the annular region, but in the domain (1.2). Thus, let

$$v(t, z, \mu) = v_0(t, z) + \mu u(t, z, \mu),$$

where

$$u(t, z, \mu) = \sum_{k=1}^{m} \mu^{k-1} v_k(t, z). \tag{7.1}$$

Moreover, let the function $R(t, z, \mu)$ be expanded in powers of μ as well:

$$R(t, z, \mu) = \sum_{k=0}^{m} \mu^k R_k(t, z).$$

Let us differentiate the perturbed Lyapunov function (7.1) by making use of equations (1.1):

$$
\begin{aligned}
\frac{dv}{dt} &= \frac{\partial v_0}{\partial t} + \frac{\partial v_0}{\partial z} f + \\
&+ \frac{\partial v_0}{\partial z} \sum_{r=0}^{m} \mu^{r+1} R_r(t, z) + \sum_{k=1}^{m} \mu^k \left(\frac{\partial v_k}{\partial t} + \frac{\partial v_k}{\partial z} f \right) + \\
&+ \sum_{j=1,i=0}^{m} \frac{\partial v_j}{\partial z} R_i(t, z) \mu^{i+j++1}.
\end{aligned} \tag{7.2}
$$

We require that the functions v_k are solutions of the recurrent system of linear first – order partial differential equations

$$\frac{\partial v_k}{\partial t} + \frac{\partial v_k}{\partial z} f = -\phi_{k-1}(t, z), \tag{7.3}$$

with

$$\phi_{k-1}(t, z) = \frac{\partial v_0}{\partial z} R_{k-1} + \sum_{j=1,i=0}^{i+j+1=k} R_i \frac{\partial v_j}{\partial z}.$$

The integral curves $z = \bar{z}(t)$ of the unperturbed system (1.3) are the characteristics of the system (7.3). Therefore, by consecutively integrating the system (7.3) with the initial values $v_k(0, z_0) = 0$, we obtain functions v_k which are equal to zero on the initial set $(t = 0, z_0)$:

$$v_k(t, z) = -\int_0^t \phi_{k-1}(t, \bar{z}(t))\, dt. \qquad (7.4)$$

Let us introduce the notation for the mean of $\phi_k(t, z)$:

$$\psi_k(t_0, z_0) = \lim_{T \to \infty} T^{-1} \int_{t_0}^{t_0+T} \phi_k(t, \bar{z}(t))\, dt. \qquad (7.5)$$

We see that if ψ_{k-1} differs from zero, then v_k is not bounded for $0 < t < \infty$. Thus, we take the number m such that $\psi_0 = \psi_1 = \ldots = \psi_{m-1} \equiv 0$, but $\psi_m \neq 0$ and assume that the solutions of Eq. (7.3) are bounded in the region (1.2) for $k \leq m$. Under these assumptions one constructs the function u, as all conditions of Theorems 2.1, 2.3 hold along with a theorem on (x, μ) – stability.

Let us formulate the main condition concerning expansion of the perturbed function $v + u$ in powers of μ: in the region (1.2) there exist bounded solutions v_k, $1 \leq k \leq m$ of Eq. (7.3), moreover, the mean ψ_m is strictly negative for $0 < \gamma <\| x_0 \|$.

2.8　Theorem on Asymptotic Stability of Perturbed Nonlinear Systems in Neutral Case

Let us consider a system of ordinary differential equations which does not contain an explicit small parameter

$$\dot{x} = f(t, x) + R(t, x) \qquad (8.1)$$

in the domain $t > 0$, $\| x \|\leq \epsilon_0$ [4].

For the sake of brevity, we investigate stability in all variables x. The position of equilibrium $x = 0$ of the truncated (perturbed) system

$$\dot{x} = f(t, x), \quad f(t, 0) \equiv 0 \qquad (8.2)$$

is stable.

The condition of asymptotic stability of the system (8.1) in terms of some restrictions on the mean can be stated as follows.

Theorem 2.6 *Suppose that in the domain $t \geq 0$, $\| x \| \leq \epsilon_0$:*

a) the vector functions $f(t, x)$ and $R(t, x)$ satisfy the Lipschitz condition with the Lipschitz constant $L > 0$ in x; moreover

$$\| R(t, x) \| \leq N \| x \|^r, \quad N > 0, \quad r > 1. \tag{8.3}$$

b) There exists a function $v(t, x)$ admitting an infinitely small upper bound and whose total derivative is non − positive by virtue of the system (8.2).

c) The function $\phi(t, x) = \nabla v \cdot R(t, x)$ is differentiable with respect to x, and there exist $M > 0$ and $d \geq r$ such that

$$|\phi(t, x)| \leq M \| x \|^d, \quad \| \nabla \phi \| \leq M \| x \|^{d-1} .$$

d) There exists a mean, negative − definite in x_0

$$\psi(t_0, x_0) = \lim_{T \to \infty} T^{-1} \int_{t_0}^{t_0+T} \phi(t, \bar{x}(t, t_0, x_0)) \, dt,$$

where $x = \bar{x}(t, t_0, x_0)$ solves the system (8.2) with the initial value $x(t_0) = x_0$ and for any $\delta_1 > 0$ there exists $\delta_2 > 0$ such that $\psi(t_0, x_0) \leq -\delta_2$ for $\| x_0 \| < \delta_1$.

e) $\| x \|^d /\psi(t, x) = O(1)$ for $\| x \| \to 0$.

f) there exists $l > 0$ such that for all $t_0 \geq 0$, $\| x_0 \| \leq \epsilon_0$, and $t - t_0 \geq l$

$$|k(t, t_0, x_0)| \leq \frac{1}{4} |\psi(t_0, x_0|, \tag{8.4}$$

where

$$|k(t, t_0, x_0)| = (t - t_0)^{-1} \int_{t_0}^{t} \phi(t, \bar{x}(t, t_0, x_0)) \, dt - \psi(t_0, x_0). \tag{8.5}$$

Then the state of equilibrium of the system (8.1) $x = 0$ is asymptotically stable.

Proof. Combining the Gronwall − Bellman lemma, condition (a) of the theorem, and the condition $f(t, 0) = 0$, we obtain the following estimates for the solutions $x(t)$ and $\bar{x}(t)$ of the systems (8.1) and (8.2)

$$\| x(t, t_0, x_0) \| \leq \| x_0 \| \exp[2nL(t - t_0)] = \| x_0 \| E_x(t - t_0), \tag{8.6}$$

$$\| x(t, t_0, x_0) \| \leq \| \bar{x}_0 \| \exp[2nL(t - t_0)] = \| \bar{x}_0 \| E_{\bar{x}}(t - t_0). \quad (8.7)$$

To estimate the difference $x(t) - \bar{x}(t)$ for the solutions emanating from the same point, we set $x_0 = \bar{x}_0$. Then

$$\| x(t) - \bar{x}(t) \| \leq$$
$$\| x_0 \|^r \frac{N}{2L} \{\exp[rnL(t - t_0)] - 1\} \exp[2nL(t - t_1)] = \quad (8.8)$$
$$\| \bar{x}_0 \|^2 E(t - t_0).$$

The functions E_x, $E_{\bar{x}}$, and E are introduced in (8.6) – (8.8) for convenience.

The inequalities (8.6) – (8.8) imply

$$|\phi(t, x(t)) - \phi(t, \bar{x}(t))| \leq \max \| \nabla v \| \| x(t) - \bar{x}(t) \| \leq$$
$$\leq \| x_0 \|^{d-1} (E_x + E_{\bar{x}})^{-1} M_0 \| x_0 \|^r E(t - t_0) = \quad (8.9)$$
$$= \| x_0 \|^{d+r-1} \Phi(t - t_0),$$

where the function Φ is defined by Eq. (8.9).

According to condition (e) of the theorem, there exists $K > 0$ such that for $t_0 \geq 0$, $\| x_0 \| \leq \epsilon_0$

$$\| x_0 \|^d \leq K |\psi(t_0, x_0)|. \quad (8.10)$$

From (8.9) and (8.10) it follows that

$$|\phi(t, x(t)) - \phi(t, \bar{x}(t))| \leq \| x_0 \|^{r-1} K \Phi(t - t_0) |\psi(t_0, x_0)|. \quad (8.11)$$

According to condition (8.3) $r > 1$, and therefore it is possible to choose ϵ_1 such that the inequalities

$$\epsilon_1^r E(2l) \leq \epsilon - \epsilon_1, \quad \epsilon_1^r K \Phi(2l) \leq \frac{1}{4}, \quad (8.12)$$

where l is the constant from condition (f) of the theorem, are satisfied. We take $w > 0$ such that $\| x_0 \| \leq \epsilon_1$ for $v(t, x) = w$. Let $x_0 \in \{x : v(t_0, x_0) \leq w\}$. By estimating the variation of $v(t, x)$ along the solution

$x = x(t, t_0, x_0)$ of the system (8.1) (as it was done in the proof of Theorem 2.1), we obtain

$$v(t, x) \leq v(t_0, x_0) + \int_{t_0}^t \phi(t, x(t))\, dt =$$

$$= v(t_0, x_0) + \int_{t_0}^t [\phi(t, x(t)) - \phi(t, \bar{x}(t, t_0, x_0))]\, dt + \qquad (8.13)$$

$$+ \int_{t_0}^t \phi(t, \bar{x}(t, t_0, x_0))\, dt.$$

Using condition (f), we now express the last integral on the RHS of (8.13) in terms of the mean. Combining the inequalities (8.4), (8.11), and (8.12) for $t_1 = t_0 + 2l$, we obtain

$$v(t_1, x(t_1)) \leq v(t_0, x(t_0)) + l\psi(t_0, x(t_0)) < v(t_0, x(t_0)) \leq w. \qquad (8.14)$$

Since all estimates are uniform with respect to t_0 and x_0, the solution $x(t)$ remains in some neighbourhood of 0 for all $t > 0$; the neighbourhood is arbitrary small for ϵ_1 small enough. Hence, the trivial solution of the system (8.1) is stable.

Now, let us take $t_k = t_0 + 2kl$, $x_k = x(t_k)$. It follows from (8.14) that $\| x \| \leq \epsilon_1$, and therefore all the estimates hold for $t = t_2$ as well, to wit

$$v(t_2, x_2) \leq v(t_0, x(t_0)) + l[\psi(t_0, x_0) + \psi(t_1, x_1)] < v(t_0, x(t_0)). \quad (8.15)$$

A similar inequality holds for any k. Thus, a series of terms of constant signs

$$\psi(t_0, x_0) + \psi(t_1, x_1) + \ldots + \psi(t_k, x_k) + \ldots \qquad (8.16)$$

converges, i.e. $\psi(t_k, x_k) \to 0$. By virtue of condition (d) of the theorem this means that $x_k \to 0$, and hence the point $x = 0$ is asymptotically stable. The theorem is proved.

2.9 Theorems on Asymptotic Stability of Standard – Form Systems and Systems with Small Perturbations

Consider a system with small parameter $\mu > 0$

$$\dot{x} = f(t, x) + \mu R(t, x). \tag{9.1}$$

Since here a small parameter is present, it is possible to weaken restrictions as compared with Theorem 2.6.

Theorem 2.7 *(cf. [5]). Suppose that the function $v(t, x)$ is differentiable twice with respect to x in the domain $t \geq 0$, $\| x \| \leq \epsilon_0$ and there exist constants $M_v > 0$ and $m \geq 1$ such that*

$$\| \nabla v \| \leq M_v \| x \|^m, \quad \left\| \frac{\partial^2 v}{\partial x^2} \right\| \leq M_v \| x \|^{m-1}.$$

Let conditions (a), (b), (d), and (f) of Theorem 2.6 be satisfied with $r \geq 1$ in (a) and $d = m + 2$ in (e).

Then there exists $\mu_0 > 0$ such that for all μ $\mu_0 \geq \mu > 0$, the equilibrium position of the system (9.1) $x = 0$ is asymptotically stable.

The proof is similar to that of Theorem 2.6.

A system in the standard form [9] is a particular case of the system (9.1) with $f(t, x) = 0$. Theorem 2.7 provides the conditions of asymptotic stability for such a system, if the equilibrium position of the averaged system is stable. This statement was proved in Ref. [104] in another fashion.

Now, let us consider a more general system where $R(t, x)$ can be represented as a sum of terms of increasing order of smallness with respect to x in the neighbourhood of $x = 0$ (cf. [2]):

$$R(t, x) = R_1(t, x) + R_2(t, x) + \dots \tag{9.2}$$

$$\| R_{j_s}(t, x) \| \leq M \| x \|^{r_j}, \quad M > 0,$$

(9.3)

$$1 < r_1 < r_2 < \ldots, \quad s = 1, \ldots, n$$

n the domain $t \geq 0$, $\| x \| \leq \epsilon_0$.

We introduce the notation

$$F_0(t, x) = (\nabla v_0) \cdot R(t, x), \quad \phi_0(t, x) = (\nabla v_0) \cdot R(t, x),$$

(9.4)

$$F_k(t, x) = \nabla(v_0 + v_1 + \ldots + v_k) \cdot R_1(t, x) -$$
$$-\phi_0 - \phi_1 - \ldots - \phi_{k-1},$$

(9.5)

where $v_s(t, x)$ are determined from the recurrent system

$$v_s(t, 0) = 0, \quad \frac{\partial v_s}{\partial t} + \nabla v_s \cdot f = -\phi_{s-1}, \quad (s = \overline{1, k})$$

(9.6)

The function $v(t, x) = v_0 + \ldots + v_k$ is called the *perturbed* Lyapunov function. It was first introduced for systems with small parameters in Ref. [60]. Here the size ϵ_0 of the domain can play a role of a small parameter.

Let us denote

$$\psi_s(t_0, y_0) = \lim_{T \to \infty} \frac{1}{T} \int_{t_0}^{t_0+T} \phi_s(t, y(t, t_0, y_0)) \, dt,$$

(9.7)

where ψ_s is the mean value of the function ϕ_s along the solution of the truncated system (9.2).

Theorem 2.8 *Suppose that in the domain $t \geq 0$, $\| x \| \leq \epsilon_0$:*

a) Functions $f_i(t, x)$ and $R_{ij}(t, x)$ satisfy the Lipschitz condition with the Lipschitz constant $L \geq 0$ in x.

b) There exists a positive – definite function $v_0(t, x)$ admitting in x an infinitely small upper bound; the derivative of v is, by virtue of the system (9.2), non – positive.

c) There exist bounded solutions v_* of the system (9.6) such that the function $v(t, x)$ is positive – definite in the vicinity of zero.

d) The function $\phi_k(t, x)$ is differentiable with respect to x and there exist constants $M > 0$ and $d > 0$ such that

$$|\phi_k(t, x)| \leq M \parallel x \parallel^d, \quad \parallel \nabla \phi_k \parallel \leq M \parallel x \parallel^{d-1}, \quad M > 0.$$

e) There exists a negative mean $\psi_k(t_0, x_0)$, uniform with respect to t_0 and x_0 and, for $k > 0$,

$$K\psi_k(t_0, x_0) \leq - \parallel x \parallel^d .$$

f) There exists $l > 0$ such that for any $t_0 \geq 0$ and $\bar{x}_0$ $(\parallel \bar{x}_0 \parallel \leq \epsilon_0)$

$$|(t - t_0)^{-1} \int_{t_0}^t \phi_k(t, \bar{x}(t, t_0, \bar{x}_0))\, dt - \psi_k(t_0, \bar{x}_0)| \leq \frac{|\psi_k(t_0, \bar{x}_0)|}{4}$$

for $t - t_0 > l$.

Then the equilibrium position $x = 0$ of the system (8.1) is asymptotically stable.

The proof is similar to that of Theorem 2.6.

2.10   Theorem on Stability of Systems Splitting without Perturbations

As we have said above, the main condition of stability in the generalized Lyapunov method is related to the sign of derivative of the mean of the generalized which is calculated by integrating along the solution of the generating system. If this mean does not exist or its sign is not fixed, the situation is becoming not determined. In such a case, it is reasonable to make use of additional considerations. One of such might be the conditions on generating system, for example the condition that the system splits into several subsystems.

Consider a system, splitting without perturbations [6]

$$\dot{x} = f(t,x) + \mu R(t,x,z), \tag{10.1}$$

$$\dot{z} = g(t,z) + \mu Q(t,x,z), \tag{10.2}$$

where dim $x = m$, dim $z = n$, the functions f, g, R, Q are defined, bounded, and satisfy the Lipschitz condition in x, z in the domain $t \geq 0$, $\| x \| + \| z \| \leq \epsilon_0$. Let us write down the equations of the unperturbed split system

$$\dot{\bar{x}} = f(t,\bar{x}), \quad \dot{\bar{z}} = g(t,\bar{z}), \tag{10.3}$$

where $f(t,0) = g(t,0) = 0$, the equilibrium position $x = z = 0$ is stable and the Lyapunov functions $v_0(t,x)$, $u_0(t,z)$ are known. We introduce the notations

$$R^0(t,x) = R(t,x,0),$$

$$\tag{10.4}$$

$$R'(t,x,z) = R(t,x,z) - R(t,x,0),$$

$$\psi_0 = \lim_{T \to \infty} \frac{1}{T} \int_{t_0}^{t_0+T} \frac{\partial v_0}{\partial x} R^0(t,\bar{x})\, dt,$$

$$\tag{10.5}$$

$$\xi_0(t_0, x_0, z_0) = \lim_{T \to \infty} \frac{1}{T} \int_{t_0}^{t_0+T} \frac{\partial u_0}{\partial z} Q(t,\bar{x},\bar{z})\, dt.$$

Theorem 2.9 *Suppose that for* t, x, z *from the domain* $t \geq 0$, $\| x \| + \| z \| \leq \epsilon_0$:

a) There exists a positive – definite function $v_0(t,x)$ *admitting in* x *an infinitely small upper bound such that*

$$\frac{\partial v_0}{\partial t} + \frac{\partial v_0}{\partial x} f(t,x) \leq 0, \quad \left\| \frac{\partial v_0}{\partial x} \right\| < \infty.$$

b) There exists a positive – definite function $u_0(t,z)$, *admitting in* z *an infinitely small upper bound such that*

$$\frac{\partial u_0}{\partial t} + \frac{\partial u_0}{\partial z} g(t,z) \leq 0.$$

c) The functions $\dfrac{\partial v_0}{\partial x} R$, $\dfrac{\partial v_0}{\partial z} Q$ satisfy the Lipschitz condition in x and z.

d) Uniformly with respect to t_0 and x_0, there exists a negative mean $\psi_0(t_0, x_0)$ satisfying $\psi_0(t_0, x_0) \leq -\chi(\| x_0 \|)$, where $\chi(\alpha) > 0$ is a non – decreasing function, $\lim\limits_{\alpha \to 0} \chi(\alpha) = 0$.

e) Uniformly with respect to t_0 x_0 and z_0, there exists a negative mean $\xi_0(t_0, x_0, z_0)$ satisfying $\xi_0(t_0, x_0, z_0) \leq -\lambda(\| z_0 \|)$, where $\lambda(\alpha) > 0$ is a non – decreasing function, $\lim\limits_{\alpha \to \infty} \lambda(\alpha) = 0$.

Then, for every $\epsilon > 0$ there exist $\delta(\epsilon)$ and $\mu_0(\epsilon)$ such that for every solution $x(t)$ and $z(t)$ of the systems (10.1) and (10.2) with the initial data satisfying the inequality $\| x_0 \| + \| z_0 \| \leq \delta$, for $0 < \mu \leq \mu_0$, $t \geq 0$, the inequality $\| x(t) \| + \| z(t) \| \leq \epsilon$ holds.

Proof. Stability in the z component follows from Theorem 2.1. All the conditions of Theorem 2.1 are satisfied because there exists the Lyapunov function $u_0(t, z)$ and condition (e) concerning the mean $\xi_0(t_0, x_0, z_0)$ is fulfilled; moreover $u(t, x, z, \mu, \epsilon) \equiv 0$. Thereafter we will need a formal formulation of this result: for every $\epsilon_1 > 0$ there exist $\delta(\epsilon_1) > 0$ and $\mu_1(\epsilon_1)$ such that for all μ, $0 < \mu < \mu_1$, and $t > 0$, the inequality $\| z(t) \| \leq \epsilon_1$ holds if $\| z_0 \| \leq \delta$.

Let us examine stability of the x component. To this end, we fix $\epsilon > 0$ ($\epsilon \leq \epsilon_0$) and choose numbers w_0, η such that the region bounded by the surface $v_0 = w_0$ is contained in the $\epsilon/2$-neighbourhood of the point $x = 0$, while for $\| x \| < \eta$ we ensure the inequality $v_0 < w_0$. Let the x component of the integral curve satisfy the condition $\| x_0 \| < \eta$ at the initial moment of time and let this curve intersect the surface $v_0 = w_0$ at the moment $t = t_0$. Consider the variation of $v_0(t, x)$ along the solution $x(t)$:

$$v_0(t, x) \leq w_0 + \mu \int_{t_0}^{t} \left[\frac{\partial v_0}{\partial x} R(t, x, z) - \frac{\partial v_0}{\partial x} R(t, \bar{x}, z) \right] dt +$$

$$(10.6)$$

$$+\mu \int_{t_0}^{t} \frac{\partial v_0}{\partial x} R^0(t, \bar{x}) \, dt + \mu \int_{t_0}^{t} \frac{\partial u}{\partial x} R'(t, \bar{x}, z) \, dt$$

As in the proof of Theorem 2.1, the first integral is estimated by making use of condition (b), the Gronwall lemma, and the Lipschitz condition. The second integral in (10.6) is estimated with the help of condition (d). Let us discuss estimation of the third integral in details. Condition (a) and (10.4) imply existence of a constant $c > 0$ such that

$$\left| \frac{\partial v_0}{\partial x} R'(t, \bar{x}, z) \right| \le \left\| \frac{\partial v_0}{\partial x} \right\| \parallel R'(t, \bar{x}, z) \parallel \le c \parallel z \parallel \le c\epsilon_1. \qquad (10.7)$$

Let ϵ_1 satisfy the inequalities

$$\epsilon_1 \le \frac{1}{2}\epsilon, \quad c\epsilon_1 \le \frac{\chi(\eta)}{6}. \qquad (10.8)$$

Along with ϵ_1, we choose appropriate $\delta > 0$ and $\mu_1 > 0$; it follows from (10.6) and (10.8) that there exist $\mu_0 < \mu_1$ and $l > 0$ such that for all $0 < \mu < \mu_0$ and $t \in [t_0, t_0 + 2l]$

$$v_0(t, x) \le w_0 + \frac{1}{2}\mu(t_0 - t)\chi(\eta) < w_0, \qquad (10.9)$$

holds starting from the moment $t > t_0 + l$.

Thus, in the coordinates $x(t)$ the curve returns to the region bounded by the surface $v_0 = w_0$. The estimates are uniform with respect to t_0, x_0 and in the coordinate x the curve remains inside the region $\parallel x(t) \parallel \le \frac{1}{2}\epsilon$ for all $t \ge t_0 + 2l$.

Now, let $\delta_0 = \min\{\delta, \eta\}$ and we consider solutions with the initial values from the δ_0-neighbourhood of zero. Then, $\parallel x(t) \parallel + \parallel z(t) \parallel \le \epsilon$ for all $t \ge 0$ and $0 < \mu \le \mu_0$. The theorem is proved.

2.11 Stability of Systems with Additional Correlations between Properties of Mean and Derivative of Lyapunov Function

Consider a system with perturbations R and small parameter μ:

$$\dot{x} = f(t, x) + \mu R(t, x), \qquad (11.1)$$

dim $x = n \geq 1$. The unperturbed system

$$\dot{x} = f(t,x), \quad f(t,0) \equiv 0 \tag{11.2}$$

has the stationary point $x = 0$ and the functions $f(t,x)$ and $R(t,x)$ are defined in the domain $t \geq 0$, $\parallel x \parallel \leq \epsilon_0$ and satisfy the condition of existence and uniqueness of solutions of the Cauchy problem. We assume that the stationary point $x = 0$ of the system (11.2) is stable, which is ensured by existence of the positive – definite Lyapunov function $v(t,x)$, having in the domain $t \geq 0$, $\parallel x \parallel \leq \epsilon_0$ derivative obeying the condition

$$\frac{\partial v}{\partial t} + (\nabla v)f(t,x) \leq J(x) \leq 0, \tag{11.3}$$

with $J(x)$ being a continuous, negative function of constant sign. Let us denote $\phi(t,x) = (\nabla v)R(t,x)$ and

$$\psi(t_0, x_0) = \lim_{T \to \infty} \frac{1}{T} \int_{t_0}^{t_0+T} \phi(t, \bar{x}(t))\, dt, \tag{11.4}$$

where $\bar{x}(t)$ is a solution of the system (11.2) with the initial data $\bar{x}(t_0) = x_0$. Moreover, let M be a set of points x such that $J(x) = 0$, $\sigma_\lambda(M)$ the λ-neighbourhood of the set M, u_η the η-neighbourhood of the point $x = 0$, and $M_\alpha = M \setminus u_\alpha$.

Theorem 2.10 *([26]) Suppose that:*

a) There exists a Lyapunov function $v(t,x)$ of the system (11.2) which is positive definite in x, has a bounded partial derivative $\dfrac{\partial v}{\partial x}$, and admits in x an infinitely small upper bound.

b) The total derivative

$$\frac{\partial v}{\partial t} + (\nabla v)f$$

satisfies the condition (11.3), where $J(x)$ is continuous and negative.

c) There exists a constant R_0 and the non – decreasing $\chi(\alpha)$ such that $\parallel R(t,x) \parallel < R_0$, $\lim\limits_{\alpha \to 0}\chi(\alpha) = 0$, and

$$|\phi(t,x') - \phi(t,x'')| < \chi(\parallel x' - x'' \parallel).$$

d) There exists a mean $\psi(t_0, x_0)$ and the limit (11.4) is uniform with respect to $t_0 > 0$.

e) There exists a function $\psi_0(x)$ which is continuous in x, negative – definite on the set M and satisfies $\psi(t, x) \leq \psi_0(x)$.

Then, for every $\epsilon > 0$ there exist $\eta > 0$ and $\mu_0 > 0$ such that any solution of the system (11.1) with $x(0) = x_0$, $\| x_0 \| < \eta$, satisfies the inequality $\| x(t) \| < \epsilon$ for $0 < \mu < \mu_0$, $t \geq 0$ (i.e. any solution is (x, μ) – stable).

Proof. Let us fix $\epsilon > 0$ and choose $0 < \epsilon_1 < \epsilon$. By virtue of condition (a), there exists a positive – definite function $w(x)$ such that $v(t, x) \geq w(x)$. We introduce the notation $\alpha = \inf_{\epsilon_1 < x < \epsilon_0} w(x)$ and fix v_0, $0 < v_0 < \alpha$. Next we consider the moving surface $v(t, x) = v_0$. Obviously, this surface lies inside the ϵ_1-neighbourhood of the point $x = 0$. According to condition (a), $v(t, x)$ admits an infinitely small upper bound. therefore, there exists η, $\eta > 0$ such that the η-neighbourhood lies inside the moving surface $\{v(t, x) = v_0\}$. Let us show that η is the value we are looking for.

By condition (e), there exist λ, $0 < \lambda < \eta$ and δ, $\delta = \delta(\eta, \lambda) > 0$ such that the mean $\psi(t_0, x_0)$ satisfies the inequality

$$\psi(t_0, x_0) < -\delta, \quad t \geq 0, \quad x_0 \in \sigma_\lambda(M_\eta). \tag{11.5}$$

Then condition (b) implies that there exists γ, $\gamma > 0$ ($\gamma = \gamma(\eta, \lambda)$) such that

$$\frac{\partial v}{\partial t} + \frac{\partial v}{\partial x} f(t, x) < -\gamma, \quad t \geq 0, \quad \| x \| > \eta, \quad x \notin \sigma_\lambda(M_\eta). \tag{11.6}$$

We represent the moving surface $\{v(t, x) = v_0\}$ in the form

$$\tilde{A} \cup \tilde{B} = \{v(t, x) = v_0\} = (\{v = v_0\} \cap \sigma_\lambda(M)) \cup (\{v = v_0\} \setminus \sigma_\lambda(M)). \tag{11.7}$$

Consider the solution $x = x(t)$ of the system (11.1) such that $\| x(0) \| < \eta$. Let this solution leave the η-neighbourhood and intersect the surface $\{v(t,x) = v_0\}$ at the point $x = x_0$ for $t = t_0$. We will show that there exists $\mu_1(\epsilon) > 0$ such that for $0 < \mu < \mu_1$, $x_0 \in \tilde{A}$, i.e. $x_0 \in \sigma_\lambda(M)$. Let as assume the contrary: $x_0 \notin \sigma_\lambda(M)$. Then from condition (b) and (11.3) we have $J(x_0) < -\gamma < 0$. Taking the expression for total derivative into consideration, we obtain

$$\frac{dv}{dt} = \frac{\partial v}{\partial t} + \frac{\partial v}{\partial x}f(t,x) + \mu\frac{\partial v}{\partial x}R(t,x) \le J(x(t)) + \mu\phi(t,x(t)). \quad (11.8)$$

Choosing μ_1 sufficiently small and taking the fact that Δv and $R(t,x)$ are bounded into account, we have

$$\frac{dv}{dt} \le -\gamma + \frac{1}{2}\gamma = -\frac{1}{2}\gamma, \quad (11.9)$$

i.e. at the moment $t = t_0$ the Lyapunov function decreases, which contradicts the assumption that the solution $x(t)$ intersects the surface $\{v(t,x) = v_0\}$ at the point $x_0 \notin \sigma_\lambda(M)$. This contradiction proves that x_0 belongs to $\sigma_\lambda(M)$ or $x_0 \in \tilde{A}$.

Then, by virtue of condition (e), the estimate (11.5) holds.

Consider evolution of the function $v(t,x)$ along the solution $x = x(t)$ of the system (11.1):

$$v(t,x(t)) \le v(t_0,x_0) + \mu \int_{t_0}^{t} \phi(t,x(t))\,dt. \quad (11.10)$$

We will estimate the integral in (11.10) as in the proof of Theorem 2.1, taking the fact that the mean is negative (condition (e)) into account. It follows that there exists μ_0, $0 < \mu_0 \le \mu_1$ such that the integral (11.10) becomes negative at some moment of time $t = t_1$, i.e. $v(t_1,x(t_1)) < v(t_0,x_0)$, and therefore, the solution $x = x(t)$ returns to the moving surface $\{v = v_0\}$. During this period of time the solution remains in the ϵ-neighbourhood. All the estimates are uniform with respect to t_0, hence the solution may leave the moving surface as described above, but after finite period of time it will return back inside this surface. The theorem is proved.

A corresponding theorem on instability which accounts for the fact that the mean may exist not in the entire neighbourhood of $x = 0$ was proved in [26].

2.12 Investigation of Stability by Averaging over Explicit Time Dependence

The theorems presented above included the assumption that the mean calculated by integrating along a solution of generating system did exist. This, in turn, made it necessary to solve the generating system itself. With respect to this, an important case where one can average over the time variable explicitly present in the equation, was studied in Ref. [25]. Consider a weakly non – autonomous system

$$\dot{x} = f(x) + \mu X(t, x), \qquad (12.1)$$

where μ is a small parameter, dim $x = n$, the functions $f(x)$ and $X(t, x)$ are defined in the domain $t \geq 0$, $\| x \| < \epsilon_0$. The generating system

$$\dot{x} = f(x), \qquad (12.2)$$

has the stable Lyapunov point $x = 0$, moreover there exists a positive – definite Lyapunov function $v(x)$ such that $(\nabla v)f(x) \leq 0$.

Let $X(t, x)$ have the mean

$$X_0(x) = \lim_{T \to \infty} \frac{1}{T} \int_{t_0}^{t_0+T} X(t, x)\, dt, \qquad (12.3)$$

such that there exists the averaged system

$$\dot{x} = f(x) + \mu X_0(x). \qquad (12.4)$$

For a system of the form (12.1) with $f(x) = 0$ the theorem on proximity of solutions of complete and averaged standard – form system on the infinite time interval is known (the averaging system [11]). Here, we will formulate conditions which guarantee the proximity for a broader class of systems that do not reduce to the standard form.

Let us first introduce the notation:

$$F(t, x) = (\nabla v)X(t, x), \qquad (12.5)$$

$$\tilde{X}(t, x) = X(t, x) - X_0(x), \qquad (12.6)$$

$$k(T, x, t_0) = \frac{1}{T} \int_{t_0}^{t_0+T} \tilde{X}(t, x)\, dt. \qquad (12.7)$$

Since the Lyapunov function $v(x)$ is positive – definite, for every $\epsilon > 0$ there exists $\eta(\epsilon) > 0$ such that the equipotential surface $\{x : v(x) = v_0\}$ lies in the annular region $\eta <\| x \|< \epsilon$.

Theorem 2.11 *Suppose that for $t \geq 0$ and $\| x \|< \epsilon_0$:*

a) There exists a positive definite differentiable Lyapunov function $v(x)$ such that $(\nabla v)f(x) \leq 0$.

b) There exist constants $M_1 > 0$ and $\beta > 0$ such that $\eta(\epsilon) \geq M_1 \epsilon^\beta$.

c) There exists the mean $X_0(x)$, defined by Eq. (12.3); the rate of convergence to the mean can be estimated as

$$|k(T, x, t_0)| \leq M_2 \| x \|^p T^{-\alpha} \quad (\lim_{T \to \infty} k = 0),$$

where $M_2 > 0$, $p > 0$, $\alpha > 0$.

d) There exist positive constants L, R_0, M_3, M_4, q, g such that

$$\| \nabla v \|\leq M_3 \| x \|^q, \quad \| \nabla F \|\leq M_4 \| x \|^g,$$

$$\| X(t, x) \|\leq R_0, \quad f(0) = 0, \quad f(x) \in Lip_x(L).$$

e) $(\nabla v)X_0(x)$ is a negative – definite function in x, $(\nabla v)X_0(x) \leq -M_5 \| x \|^d$, with $M_5 > 0$, $d > 0$.

f) $f(x)$ satisfies the condition

$$\| f(x) \|\leq M_0 \| x \|^\sigma,$$

where

$$M_0 > 0, \quad \sigma > \beta d(1 + \frac{1}{\alpha}) - \frac{1}{\alpha}(p + q + \alpha g).$$

Then for each $\epsilon > 0$ there exist $\eta_0(\epsilon) > 0$ and $\mu_0(\epsilon) > 0$ such that for $0 < \mu < \mu_0(\epsilon)$ every solution $x(t)$ of the system (12.1) with $\| x(0) \| < \eta_0(\epsilon)$ satisfies the condition $\| x(t) \| < \epsilon$ for $t \geq T > 0$.

Proof. Let us fix $\epsilon > 0$ ($\epsilon \leq \epsilon_0$) and set $\eta_0 = \eta(\epsilon/m)$, where $m \geq 2$ will be chosen later. Consider an arbitrary solution $x = x(t)$ for which $\| x(0) \| \leq \eta_0$. Suppose that $x = x(t)$ leaves the η_0-neighbourhood of the origin at $t = t_0$ and moves outside the equipotential surface $\{x : v(x) = v_0\}$, belonging to the annular region $\eta(\epsilon/m) \leq \| x \| \leq \epsilon/m$.

Consider evolution of the Lyapunov function $v(x)$ along the solution $x(t)$ of the system (12.1) for $t \geq t_0$. By virtue of condition (a)

$$v(x(t)) \leq v(x_0) + \mu \int_{t_0}^{t} (\nabla v) X(t, x(t)) \, dt. \tag{12.8}$$

Let us show that at $t = t_1(\epsilon)$ the integral in (12.8) becomes negative, i.e., for sufficiently small μ and sufficiently large $m \geq 2$, $x(t)$ returns inside the equipotential surface $v(x) = v_0$ without leaving the ϵ-neighbourhood. To this end, we represent the integral as

$$I = \int_{t_0}^{t} F(t, x_0) \, dt + \int_{t_0}^{t} [F(t, x(t)) - F(t, \bar{x}(t))] \, dt +$$

$$\tag{12.9}$$

$$+ \int_{t_0}^{t} [F(t, \bar{x}(t)) - F(t, x_0)] \, dt = I_1 + I_2 + I_3,$$

where $x = \bar{x}(t)$ is a solution of the autonomous system (12.2) with the initial data $\bar{x}(t_0) = x_0 = x(t_0)$.

By virtue of condition (c)

$$I_1 = \int_{t_0}^{t} F(t, x_0) \, dt =$$
$$(t - t_0)[(\nabla v(x)) X(x_0) + (\nabla v(x_0)) k(t - t_0, x_0, t)]. \tag{12.10}$$

From conditions (b) and (e), we have

$$(\nabla v(x))X_0(x) \le -M_5 \parallel x \parallel^d \le -M_5\eta^d\left(\frac{\epsilon}{m}\right) = -\delta.$$

It follows from conditions (c) and (d) that

$$\parallel (\nabla v)\cdot k \parallel \le \parallel \nabla v \parallel \parallel k \parallel \le$$

$$\le M_3 \parallel x_0 \parallel^q M_2 \parallel x_0 \parallel^p (t - t_0)^{-\alpha} \le M_3 M_2 \left(\frac{\epsilon}{m}\right)^{q+p} (t - t_0)^{-\alpha}.$$

We choose $t \ge t_0 + l(\epsilon/m)$ such that

$$\parallel (\nabla v)\cdot k \parallel \le \frac{\delta}{4}.$$

To determine $l(\epsilon/m)$, we make use of the equation

$$M_3 M_2 \left(\frac{\epsilon}{m}\right)^{q+p} l^{-\alpha} = \frac{1}{4}M_5 M_1^d \left(\frac{\epsilon}{m}\right)^{\beta d},$$

hence,

$$l\left(\frac{\epsilon}{m}\right) = c_1 \left(\frac{\epsilon}{m}\right)^{p_1}, \quad c_1 = \left(\frac{4M_2 M_3}{M_5 M_1^d}\right)^{\frac{1}{\alpha}} > 0,$$

$$p_1 = \frac{q + p - \beta d}{\alpha}. \tag{12.11}$$

For $t > t_0 + l$, we have the estimate

$$I_1 \le (t - t_0)[-\delta + \frac{1}{4}\delta] = -\frac{3}{4}(t - t_0)\delta. \tag{12.12}$$

Finally, we choose $\mu_1(\epsilon/m)$ such that for $\mu < \mu_1$

$$\parallel x(t) - \bar{x}(t) \parallel \le \mu c_1 \le \frac{1}{2}\epsilon \le \epsilon - \frac{\epsilon}{m},$$

and then, since the stationary point $x = 0$ of the generating system (12.2) is stable, we obtain

$$\parallel x(t) \parallel \le \parallel \bar{x}(t) \parallel + \parallel x(t) - \bar{x}(t) \parallel \le \frac{\epsilon}{m} + \left(\epsilon - \frac{\epsilon}{m}\right) = \epsilon,$$

i.e. the solution $x = x(t)$ does not leave the ϵ-neighbourhood of the point $x = 0$ for $t_0 \le t \le t_0 + 2l$ and the estimate $\| x(t) - \bar{x}(t) \| \le \mu C_1$ holds on this interval.

Let us estimate the integral I_2 for $t_0 \le t \le t_0 + 2l$. By virtue of condition (d)

$$|I_2| = | \int_{t_0}^t [F(t, x(t)) - F(t, \bar{x}(t))] \, dt | \le$$

$$\le (t - t_0) \max_{t \in [t_0, t_0 + 2l]} \| \nabla F \| \max_{t \in [t_0, t_0 + 2l]} \| x(t) - \bar{x}(t) \| \le$$

$$\le (t - t_0) D_1 \mu.$$

We choose μ_2, $\mu_2 < \mu_1$, such that for $0 < \mu < \mu_2$

$$|I_2| \le (t - t_0) D_1 \mu < \frac{1}{4}(t - t_0)\delta. \tag{12.13}$$

Since the stationary point of the system (12.2) is stable,

$$\| \bar{x}(t) - x_0 \| \le (t - t_0) \max_{t \in [t_0, t_0 + 2l]} \| f(\bar{x}(t) \| \le$$

$$2l \max_{\|x\| \le (\epsilon/m)} \| f(x) \| .$$

Using this last inequality, we obtain the estimate for I_3

$$|I_3| = | \int_{t_0}^t [F(t, \bar{x}(t)) - F(t, x_0)] \, dt | \le$$

$$\le (t - t_0) \max_{\|x\| \le (\epsilon/m)} \| \nabla F \| 2l \max_{\|x\| \le (\epsilon/m)} \| f(x) \| .$$

Combining this estimate with conditions (d), (f) and Eq. (12.11), we get

$$|I_3| \le (t - t_0) M_4 \left(\frac{\epsilon}{m}\right)^q 2l M_0 \left(\frac{\epsilon}{m}\right)^\gamma \le (t - t_0) c_2 \left(\frac{\epsilon}{m}\right)^{p_2}, \tag{12.14}$$

where

$$c_2 = 2c_1 M_0 M_4, \quad p_2 = g + \sigma + p_1.$$

We choose m sufficiently large to ensure the inequality

$$|I_3| \le (t - t_0) c_2 \left(\frac{\epsilon}{m}\right)^{p_2} \le$$

$$\le \frac{1}{4}(t - t_0) M_5 M_1^d \left(\frac{\epsilon}{m}\right)^{\beta d} = (t - t_0)\frac{\delta}{4} \tag{12.15}$$

for $t_0 \leq t \leq t_0 + 2l$.Condition (f) implies that $p_2 > \beta d$. Then, $\epsilon^{p_2} < \epsilon^{\beta d} < 1$ and to ensure (12.15) it is sufficient to take $m \geq 2$, to wit,

$$m \geq \left(\frac{4c_2}{M_5 M_1^d} \right)^{1/(p_2 - \beta d)}.$$

Combining the estimates for three integrals (12.12), (12.13), and (12.15), we obtain the required estimate for the integral I in (12.7) for $0 < \mu < \mu_0(\epsilon) \leq \mu_2(\epsilon/m)$ with $t_0 + l \leq t \leq t_0 + 2l$. Thus, the solution $x = x(t)$ can leave the surface $\{v(x) = v_0\}$ an infinite number of times, but the solution will always remain in the ϵ-neighbourhood, i.e. $x = 0$ is (x, μ) – stable.

2.13 Investigation of Stability by Averaging along Solutions of Linear System

Let us consider a system of the form

$$\dot{x} = Ax + P(t, x) + \mu R(t, x), \tag{13.1}$$

for $t \geq 0$, $\parallel x \parallel \leq \epsilon_0$, where A is a constant $n \times n$ matrix and with a generic form of nonlinearity of $P(t, x)$, $\parallel P(t, x) \parallel \leq \parallel x \parallel^\sigma$. Below, we will impose some restrictions on σ, in order to be able to carry out averaging along a solution of the linear system

$$\dot{y} = Ay. \tag{13.2}$$

With regards to the unperturbed system

$$\dot{x} = Ax + P(t, x) = f(t, x), \tag{13.3}$$

we assume that there exists a Lyapunov function $v(t, x)$ which is positive – definite, admits an infinitely small upper bound in x, and has a non – -positive derivative

$$\frac{\partial v}{\partial t} + (\nabla v) f(t, x) \leq 0. \tag{13.4}$$

Then, for every $\epsilon > 0$ there exists $\eta(\epsilon) > 0$ such that the moving equipotential surface $v(t, x) = v_0$ lies in the annular region $\eta(\epsilon) \leq \| x \| \leq \epsilon$.

We assume that the mean

$$\psi(t_0, x_0) = \lim_{T \to \infty} \frac{1}{T} \int_{t_0}^{t_0+T} (\nabla v) \cdot R(t, y)\, dt \qquad (13.5)$$

exists and can be calculated along a solution of the linear system (13.2) with the initial data $y(t_0) = x_0$. Let us introduce the function

$$\phi(t, x) = (\nabla v) \cdot R(t, x),$$
$$k(T, t_0, x_0) = \frac{1}{T} \int_{t_0}^{t_0+T} \phi(t, y(t))\, dt - \psi(t_0, x_0). \qquad (13.6)$$

Theorem 2.12 *Suppose that:*

a) There exists a positive – definite Lyapunov function $v(t, x)$ admitting an infinitely small upper bound for which

$$\frac{\partial v}{\partial t} + (\nabla v) f(x) \leq 0.$$

b) There exist constants $M_1 > 0$ and $\beta > 0$ such that $\eta(\epsilon) \geq M_1 \epsilon^\beta$.

c) There exists the mean (13.5) and the estimate

$$\psi(t_0, x_0) \leq -M_2 \| x_0 \|^d, \quad M_2 > 0, \quad d > 0$$

holds.

d) The rate of convergence to the mean is estimated as

$$|k(T, t_0, x_0)| \leq M_3 \| x_0 \|^p\, T^{-\alpha} \quad M_3 > 0, \quad p, \alpha > 0.$$

e) The linear system (13.2) possesses Lyapunov stability.

f) There exist positive constants $L, R_0, M_4, M_5, g, \sigma$ such that

$$\left\| \frac{\partial v}{\partial x} \right\| \leq M_4 \| x \|^g, \quad \| P(t, x) \| \leq M_5 \| x \|^\sigma,$$
$$\sigma > bd(1 + \alpha^{-1}) - (p + \alpha g)\alpha^{-1}$$
$$\| R(t, x) \| \leq R_0, \quad f(t, 0) = 0, \quad f(t, x) \in Lip_x(L).$$

Then for each $\epsilon > 0$ there exist $\eta(\epsilon) > 0$ and $\mu_0(\epsilon) > 0$ such that if $\parallel x(0) \parallel < \eta(\epsilon)$ and $0 < \mu < \mu_0(\epsilon)$, then $\parallel x(t) \parallel < \epsilon$ for all $t \geq 0$, i.e. the solution of the system (13.1) is (x, μ) – stable.

Proof of this theorem is similar to that of Theorem 2.10.

Theorems 2.10 and 2.11 can be used to investigate stability of systems with a weak quasi – periodic non – autonomy [25] of the form

$$\dot{x} = f(x) + \mu X(\omega t, x), \qquad (13.7)$$

where $\omega = (\omega_1, \ldots, \omega_m)$ is a constant frequency vector, while $X(\phi, x)$ is a function which is 2π – periodic with respect to $\phi = (\phi_1, \ldots, \phi_m)$. Since the frequencies are expected to be not commensurable, the system (13.7) is related to the more general class of systems called the multi – frequency systems

$$\dot{x} = f(x) + \mu X(\phi, x, \mu), \quad \dot{\phi} = \omega + \mu \Phi(\phi, x, \mu). \qquad (13.8)$$

For the system (13.8), the questions of existence of a quasi – periodic solution, the behaviour of the general solution in the neighbourhood of the quasi – periodic solution and other questions were studied in [11] and [12]. It should be stressed that these studies were undertaken either for $f(x) \equiv 0$, i.e. for standard – form systems, or for $f(x) = Px$, i.e. in linear case. The essentially nonlinear case when $\parallel f(x) \parallel \leq C \parallel x \parallel^\sigma$, $\sigma > 1$, was considered in Ref. [25]. In this work, in spite of the nonlinearity condition, the linear dependence $\eta(\epsilon) \sim C\epsilon$ was assumed, the condition of strong incommensurability of frequencies ω and the corresponding constraints on the rate of convergence of Fourier series was imposed, and the orders of nonlinearity of the functions v, ∇v, $X_0(x)$ were restricted.

2.14 Investigation of Stability of Integro – Differential Systems

1. Let us consider the system of integro – differential equations [46]

$$\dot{x} = f(t,x) + \mu X(t,x, \int_0^t F(t,s,x(s))\,ds), \qquad (14.1)$$

where μ, $0 < \mu \leq \mu_0$, is a small parameter, the functions $f(t,x)$, $X(t,x,y)$, $F(t,s,x)$ are defined for $t \geq 0$, $s \geq 0$, $\| x \| \leq \epsilon_0$, $\| y \| \leq \bar{\epsilon}_0$, dim $y = m$. The generating system

$$\dot{x} = f(t,x), \quad f(t,0) = 0 \qquad (14.2)$$

has a stable equilibrium position $x = 0$ and there exists a Lyapunov function $v(t,x)$.

We will examine stability of the system (14.1) with the help of averaging method. Averaging in integro – differential systems was discussed in Refs. [46], [63], [88], [27]. Various averaging schemes were suggested for finite and infinite time intervals. In what follows, we extend the ideas of the generalized Lyapunov method to the class of systems of the form (14.1) and suggest new schemes of averaging.

Let us introduce the notation

$$\phi(t,x,y) = \frac{\partial v}{\partial x} \cdot X(t,x,y) \qquad (14.3)$$

and define the mean value of ϕ as follows

$$\psi_1(t_0,x_0) = \lim_{T \to \infty} \frac{1}{T} \int_{t_0}^{t_0+T} \phi\left(\tau, \bar{x}(\tau), \int_0^\tau F(\tau,s,\bar{\bar{x}}(s))\,ds\right) d\tau. \qquad (14.4)$$

Here $x(t)$ is a solution of the complete system, $x(t_0) = x_0$, $\bar{x}(t)$ is a solution of the generating system of equations, $\bar{x}(t_0) = x(t_0) = x_0$, $\bar{\bar{x}}(t) = x(t)$ for $0 \leq t \leq t_0$, $\bar{\bar{x}}(t) = \bar{x}(t)$ for $t_0 \leq t < \infty$.

Similarly, we define

$$\psi_2(t_0,x_0) = \lim_{T \to \infty} \frac{1}{T} \int_{t_0}^{t_0+T} \phi\left(\tau, \bar{x}(\tau), \int_0^\tau F(\tau,s,\bar{\bar{x}}(s))\,ds\right) d\tau, \qquad (14.5)$$

$$\bar{\bar{x}}(t) = x_0, \ 0 \leq t \leq t_0, \ \bar{\bar{x}}(t) = \bar{x}(t), \ t_0 \leq t < \infty.$$

Theorem 2.13 *Suppose that in the region $t \geq 0$, $s \geq 0$, $\| x \| \leq \epsilon_0$, $\| y \| \leq \bar{\epsilon}_0$:*

a) There exists a positive – definite Lyapunov function $v(t, x)$ admitting an infinitely small upper bound in x, and satisfying in this variable the Lipschitz condition with the Lipschitz constant L_1.

b) There exist constants $M_2 > 0$, $L_2 > 0$, and $L_3 > 0$ such that $\| X(t, x, y) \| \leq M_2$, the function $X(t, x, y)$ satisfies the Lipschitz condition in variables x and y with the Lipschitz constant L_2, and the function $f(t, x)$ satisfies the Lipschitz condition in the variables x and $f(t, 0) = 0$.

c) Uniformly with respect to t_0, x_0 there exist the average $\psi_1(t_0, x_0)$; for every $\eta > 0$ ($\eta < \epsilon_0$) and $t_0 > 0$ there exists $\delta(\eta) > 0$ such that $\psi_1(t_0, x_0) < -\delta$ if $\| x_0 \| > \eta$.

d) The function $F(t, s, x)$ satisfies the Lipschitz condition with the Lipschitz constant L_4 and there exists a function $F(t, s)$, integrable over $0 \leq s < \infty$, such that

$$\| F(t, s, x) \| \leq \bar{F}(t, s), \qquad \int_0^t \bar{F}(t, s) \, ds \leq \bar{\epsilon}_0.$$

Then for every $\epsilon > 0$ there exist $\eta(\epsilon) > 0$ and $\mu_0(\epsilon) > 0$ such that if $0 < \mu < \mu_0(\epsilon)$, then every solution $x(t)$ of the system (14.1) satisfies the inequality $\| x(t) \| < \epsilon$ for $t \geq 0$, provided that $\| x(0) \| < \eta(\epsilon)$.

Proof. The theorem is proved according to the scheme employed in the generalized Lyapunov second method. Let us fix ϵ ($\epsilon < \epsilon_0$) and set $\epsilon_1 = \epsilon/2$. By condition (a) there exist quantities $v_{\epsilon_1} > 0$ and $\eta(\epsilon_1)$ such that the moving surface of equipotential of the equipotential line $\{ x : v(t, x) = v_{\epsilon_1} \}$ lies entirely within the annular region $\eta \leq \| x \| \leq \epsilon$ at $t > 0$.

We consider an arbitrary solution $x = x(t)$ of the system (14.1) for which $\| x(0) \| \leq \eta$. Assume that at the certain moment $t = t_0$, the solution $x = x(t)$ intersects the moving surface $\{ x : v(t, x) = v_\epsilon \}$ at the point $x(t_0) = x_0$. For $t \geq t_0$, by virtue of condition (a), we have

$$v(t, x(t)) \leq v(t_0, x_0) + \mu \int_{t_0}^{t_0+T} \phi(\tau, x(\tau), \int_0^\tau F(\tau, s, x(s)) \, ds) \, d\tau \quad (14.6)$$

Let us show that there exists a moment $t = t_1$ when the integral I in (14.6) becomes negative and that the solution remains in the ϵ-neighbourhood during $t_0 \leq t \leq t_1$, i.e. the conclusion of the theorem holds.

Let us rewrite the integral I in (14.6) as a sum of three integrals

$$I = I_1 + I_2 + I_3, \quad (14.7)$$

where

$$I_1(t) = \int_{t_0}^t \phi(\tau, \bar{x}(\tau), \int_0^\tau F(\tau, s, \bar{\bar{x}}(s)) \, ds) \, d\tau,$$

$$I_2(t) = \int_{t_0}^t [\phi(\tau, x(\tau), \int_0^\tau F(\tau, s, x(s)) \, ds) -$$
$$-\phi(\tau, \bar{x}(\tau), \int_0^\tau F(\tau, s, x(s)) \, ds)] \, d\tau,$$

$$I_3(t) = \int_{t_0}^t [\phi(\tau, \bar{x}(\tau), \int_0^\tau F(\tau, s, x(s)) \, ds) -$$
$$-\phi(\tau, \bar{x}(\tau), \int_0^\tau F(\tau, s, \bar{\bar{x}}(s)) \, ds)] \, d\tau.$$

By virtue of condition (c), for $t \geq t_0 + l$,

$$I_1(t) \leq -(t - t_0)\frac{3}{4}\delta, \quad (14.8)$$

where $\delta = \delta(\eta)$, $l = l(\delta)$. For $t_0 \leq t \leq t_0 + 2l$, conditions (b) and (d) imply

$$\| x(t) - \bar{x}(t) \| \leq \mu C_1. \quad (14.9)$$

Hence, for $t_0 \leq t \leq t_0 + 2l$, $0 < \mu \leq \mu_0(\epsilon)$ we obtain the estimates

$$\| x(t) \| \leq \| \bar{x}(t) \| + \| x(t) - \bar{x}(t) \| \leq \epsilon,$$

$$|I_2(t)| \leq (t - t_0)\frac{\delta}{4}, \quad (14.10)$$

$$|I_3(t)| \leq (t - t_0)\frac{\delta}{4}.$$

Combining Eqs. (14.7) and (14.8), we prove that the integral I is negative and conclusion of the theorem follows.

Employing definition of the mean (14.5) we can now formulate conditions of stability.

Theorem 2.14 *Suppose that conditions (a) and (b) of Theorem* 2.12 *are satisfied and in addition*

c) *Uniformly with respect to* t_0 *and* x_0 *there exists the mean* $\psi_2(t_0, x_0)$ *which is negative – definite in the variable* x_0.

d) $F(\tau, s, x) \in Lip_x(L_4)$ *and there exist functions* $\bar{F}(\tau, s)$ *and* $p(\tau, T)$ *such that*

$$|F(\tau, s, x)| \le \bar{F}(\tau, s),$$

$$p(\tau, T) = \int_0^T \bar{F}(\tau, s)\, ds < \bar{\epsilon}_0.$$

e) *There exists a function* $p_0(l)$, $p_0(l) \to 0$ *s* $l \to 0$ *such that*

$$\int_T^{T+l} p(\tau, T)\, d\tau \le l p_0(l).$$

Then the conclusion of Theorem 2.12 *holds, i.e. the system* (14.1) *is* (x, μ) *– stable.*

The proof of this theorem is similar to that of Theorem 2.12.

2. When $f(t, x) \equiv 0$, the system (14.1) acquires the standard form

$$\dot{x} = \mu X(t, x, \int_0^t F(t, s, x(s))\, ds). \tag{14.11}$$

Several averaging schemes were studied in connection with this system ([63], [88], [27]). Following one of them, let us average the system (14.9) over the time variable, which enters the system explicitly:

$$X_0(x) = \lim_{T \to \infty} \frac{1}{T} \int_{t_0}^{t_0+T} X(t, x, \int_0^t F(t, s, x)\, ds)\, dt. \tag{14.12}$$

Let us assume that the averaged system

$$\dot{\xi} = \mu X_0(\xi) \tag{14.13}$$

has the asymptotically stable stationary point $\xi = 0$ and that there exists a positive – definite Lyapunov function $v(x)$ such that $(\nabla v)X_0(x)$ is negative – definite. Let us further assume that the functions $X(t, x, y)$, $F(t, x, y)$ fulfill conditions of Theorem 2.12. Then this theorem implies that the point $x = 0$ is (x, μ) – stable.

3. Now, let us use the theorems proved above to study stability of the system of integro – differential equations with the kernel of difference – type:

$$\dot{x} = f(t, x) + \mu X(t, x, \int_0^t K(t - s)\Phi(x(s)) \, ds). \quad (14.14)$$

We assume that the unperturbed system has the Lyapunov – stable stationary point $x = 0$ and that there exists a corresponding Lyapunov function; the functions $f(t, x)$ and $X(t, x, y)$ satisfy conditions of Theorem 2.12.

The system (14.14) describes many problems in the theory of visco – elasticity. The kernel $K(t) > 0$ tends to 0 as t grows and is integrable for $0 \le t < \infty$.

Let us examine the mean

$$\psi_0(t_0, x_0) = \lim_{T \to \infty} \frac{1}{T} \int_{t_0}^{t_0+T} \phi(\tau, \bar{x}(\tau), \int_0^\tau K(t-s)\Phi(\bar{x}(s)) \, ds) d\tau, \quad (14.15)$$

where $\bar{x}(t)$ is a solution of the system

$$\dot{\bar{x}} = f(t, \bar{x}), \quad \bar{x}(t_0) = x_0.$$

Theorem 2.15 *Suppose that conditions (a) and (b) of Theorem 2.12 are satisfied and, moreover:*

c) Uniformly with respect to t_0 and x_0 there exists the mean $\psi_0(t_0, x_0)$ defined by Eq. (14.15), being a negative – definite function of x_0.

d) $\| \phi(x) \| \le \Phi_0$, $\Phi(x) \in Lip_x(\lambda)$, $K(t) > 0$, $\int_0^\infty K(t) \, dt = K_0$, where $\Phi_0 > 0$, $\lambda > 0$, $K_0 > 0$, $\Phi_0 K_0 \le \bar{\varepsilon}$.

Then the assertion of Theorem 2.12 is valid.

Proof. To prove this theorem, we will show that conditions of Theorems 2.12 and 2.13 are satisfied.

Using conditions of Theorem 2.14 to calculate the mean $\psi_0(t_0, x_0)$ according to the algorithm (14.2), we obtain (14.15). In our case $F(\tau, s, x) = K(\tau - s)\Phi(x)$, $\bar{F}(\tau, s) = K(\tau - s)\Phi_0$, and

$$\int_0^\tau \bar{F}(\tau, s) \, ds = \Phi_0 \int_0^\tau K(s) \, ds \leq \Phi_0 K_0 \leq \bar{\epsilon}_0. \qquad (14.16)$$

Therefore, we see that all conditions of Theorem 2.12 are fulfilled. Similarly one can check conditions of Theorem 2.13.

Remark. Using more sophisticated estimates, we can conclude from Theorems 2.12 and 2.13 that the solution $x(t), t > T(\mu)$ of the system (14.1) does not leave the $\lambda(\mu)$-neighbourhood of the origin, where $\lambda(\mu) \to 0$ as $\mu \to 0$. This can be done as in the proof of Theorem 2.1. Moreover, theorems on instability hold for the system (14.1).

2.15   On Numerical Realization of Theorems of Generalized Lyapunov Second Method

The theorems pertaining to the generalized Lyapunov second method contain as one of conditions the requirement that the generalized Lyapunov function is negative – definite. It is important to be able to check this condition by numerical computer integration.

Consider the system

$$\dot{z} = f(z) + \mu R(z), \qquad (15.1)$$

where $z = (x, y)$, μ is a small parameter, $\dim x = N$, $\dim y = M$, $\| x \| \leq H$, $f(x, y) = f(x, y + 2\pi)$, $R(x, y) = R(x, y + 2\pi)$, $f(z) = (X(z), Y(z))$, $R(z) = (R_1(z), R_2(z))$.

a) Let the function $f(z)$ be represented as a sum of two vector functions $f(z) = (X(z), Y(z)) = (X_0(z), Y_0(z)) + (X_1(z), Y_1(z))$. A structure of the individual terms is determined by the following condition:

for every $a \geq 1$

$$X_0(ax, y) = a^n X_0(x, y), \quad Y_0(ax, y) = a^{n-1} Y_0(x, y), \quad n \geq 1,$$

there exist $L_1(x) \geq 0$, $L_2(x) \geq 0$, $p \geq 0$ such that

$$\| X_1(x, y) \| \leq L_1(x), \quad L_1(ax) \geq a^{n+p} L_1(x);$$

and

$$\| Y_1(x, y) \| \leq L_2(x), \quad L_2(ax) \geq a^{n+p-1} L_2(x).$$

b) Let the structure of the functions $R(z)$ be determined by the following condition: there exist $r_1(x)$ and $r_2(x)$, and $s > 1$ such that for any $a \geq 1$

$$\| R_1(x, y) \| \leq r_1(x), \quad r_1(ax) \geq a^s r_1(x);$$

and

$$\| R_2(x, y) \| \leq r_2(x), \quad r_2(ax) \geq a^{s-1} r_2(x).$$

Let us consider the unperturbed system

$$\dot{z} = f(z). \tag{15.2}$$

c) Let there exist a Lyapunov function $v(z)$ positive – definite in x with

$$\dot{v} = \nabla v\, f(z) \leq 0.$$

By virtue of the system (15.1), we can write down the time derivative of v

$$\dot{v} = \frac{\partial v}{\partial z} f(z) + \mu \frac{\partial v}{\partial z} R(z). \tag{15.3}$$

d) Let the function $\phi(z) = \nabla v\, R(z)$ can be represented as a sum $\phi(z) = \phi_0 + \phi_1$, where ϕ_0 and ϕ_1 have different orders, to wit, for every $a \geq 1$ there exists $m \geq 1$ such that

$$\phi_0(ax, y) = a^m \phi_0(x, y),$$

and there exist $\beta(x) > 0$ and $q > 0$ such that

$$|\phi_1(x,y)| \le \beta(x), \quad \beta(ax) \ge a^{m+q}\beta(x).$$

Let us consider the system

$$\dot{z} = f_0(z), \quad f_0(z) = (X_0(z), Y_0(z)). \tag{15.4}$$

e) Let the functions X_0, Y_0, L_1, L_2, r_1, r_2 satisfy the Lipschitz condition with the Lipschitz constant N and let the functions f_0 and ϕ_0 be periodic in y with the period 2π.

f) Let there exist $T_0 > 0$ and $\delta_1(T_0)$ such that

$$\max_{0 \le \bar{\bar{y}}_0(0) \le 2\pi} \max_{\|\bar{\bar{x}}_0(0)\|=1} \min_{0 \le t \le T_0} \int_0^t \phi_0(\bar{\bar{x}}(t), \bar{\bar{y}}(t))\, dt = -\delta_1(T_0), \tag{15.5}$$

where $(\bar{\bar{x}}(t), \bar{\bar{y}}(t))$ is a solution of the system (15.4).

It should be noted that condition (f) with various initial data $\bar{x}_0(0)$, $\bar{y}_0(0)$ and upper integration limit t can be checked by numerical integration.

Theorem 2.16 *Let conditions (a) to (f) be fulfilled. If $s > n$, then for $\mu \in [\alpha, \beta]$ (α, β are finite), the equilibrium position is asymptotically stable with respect to x.*

If $s = n$, then there exists $\mu_0 > 0$ such that for $0 < \mu \le \mu_0$ the equilibrium position is asymptotically stable with respect to x.

Proof. Let us fix a number v_0, so that the moving surfaces $v(x,y) = v_0$ lie inside spheres of radii 1 and H and the solution of the system (15.1) starting from the surface $v(x,y) = v_0$ remains, in the coordinates x, in the region $\| x \| < H$ during the time T_0. Let us construct the solutions $z = z_1(t)$ of the system (15.1) and $z = \bar{\bar{z}}_1(t)$ of the system (15.4), both starting from the same point of the surface $v(x,y) = v(x_1, y_1)$. In addition, we must choose $a > 0$ and v_0 such that $\| ax_1 \| = 1$. We also construct the integral curve $z = \bar{\bar{z}}_0(t)$ of the system (15.4) from the point (ax_1, y_1). By virtue of condition (a), the solutions $z_1(t)$, $\bar{\bar{z}}_1(t)$, and $\bar{\bar{z}}_0(t)$ are related as follows:

$$\{a\bar{\bar{x}}(a^{n-1}t), \bar{\bar{y}}(a^{n-1}t)\} = \{\bar{\bar{x}}_a(t), \bar{\bar{y}}_a(t)\};\tag{15.6}$$

moreover, $a\bar{\bar{x}}(0) = \bar{\bar{x}}_a(0)$, $\bar{\bar{y}}(0) = \bar{\bar{y}}_a(0)$.

Let us introduce the constant $M_1(T_0)$:

$$M_1(T_0) = \max_{0 \leq \bar{\bar{y}}(0) \leq 2\pi} \max_{\|\bar{\bar{x}}(0)\| = 1} \max_{0 \leq t \leq T_0} (\| x \| + \| y \|).$$

Then, by Eq. (15.6), we get

$$\| \bar{\bar{x}}(a^{n-1}t) \| = a^{-1} \| \bar{\bar{x}}_a(t) \| \leq a^{-1} M_1,$$
$$\| \bar{\bar{y}}(a^{n-1}t) \| = \| \bar{\bar{y}}_a(t) \| \leq M_1,\tag{15.7}$$
$$0 \leq t \leq T.$$

To estimate the difference between the solutions $z = z_1(t)$ and $z = \bar{\bar{z}}_1(t)$, we use conditions (a), (b), (e):

$$\| ax_1(t) - a\bar{\bar{x}}_1(t) \| \leq$$
$$\leq a \int_0^t \{\| X_0(z_1) - X_0(\bar{\bar{z}}_1(t)) \| + L_1(x_1) + \mu r_1(x_1)\} \, dt \leq\tag{15.8}$$
$$a^{-n+1} \int_0^t N(1 + a^{-p} + \mu a^{-s+n})(\| ax_1 - a\bar{\bar{x}}_1(t) \| +$$
$$\| y_1 - \bar{\bar{y}}_1(t) \|) \, dt + a^{-n+1}[L_1(M_1)a^{-p} + \mu a^{-s+n} r_1(M_1)]a^{n-1}T_0.$$

The difference $\| y_1(t) - \bar{\bar{y}}_1(t) \|$ can be estimated similarly. Next, applying the Gronwall lemma, we obtain

$$\| x_1(t) - \bar{\bar{x}}_1(t) \| \leq T_0[a^{-p} \sum_{i=1}^{2} L_i(M_1) +$$

$$\tag{15.9}$$

$$+ \mu a^{-s+n} \sum_{i=1}^{2} r_i(M_1)](e^{NT_0(2+\mu)} - 1)a^{-1} \leq a^{-1-k} \tilde{M}_2,$$

where

$$\tilde{M}_2 = T_0[\sum_{i=1}^{2} L_i(M_1) + \mu \sum_{i=1}^{2} r_i(M_1)](e^{NT_0(2+\mu)} - 1), \quad (15.10)$$

$$k = \min(p, s - n) \geq 0.$$

Taking Eq. (15.7) into account, we estimate the solution from above

$$\parallel x_1(t) \parallel \leq a^{-1}M_1 + a^{-1-k}\tilde{M}_2 \leq (M_1 + \tilde{M}_2)a^{-1} = M_2 a^{-1},$$

$$\parallel y_1(t) \parallel \leq M_2, \quad\quad\quad\quad (15.11)$$

$$0 \leq t \leq a^{n-1}T_0.$$

Let us turn now to estimation of derivative of Lyapunov function:

$$v \leq v_0 + \mu \int_0^t \phi_0(\bar{\bar{z}}_1(t))\, dt + $$

$$\quad\quad\quad\quad\quad\quad\quad\quad\quad (15.12)$$

$$+\mu \int_0^t \phi_1(z_1(t))\, dt + \mu \int_0^t [\phi_0(z_1(t)) - \phi_0(\bar{\bar{z}}_1(t))]\, dt.$$

For the first integral in (15.12), by combining Eq. (15.6) and conditions (d), (f), we find t^*, $0 \leq t^* \leq T_0$ such that

$$\int_0^{t^*} \phi_0(\bar{\bar{z}}_1(t))\, dt = \int_0^{t^*} \phi_0(a\bar{\bar{x}}_1(a^{n-1}t), \bar{\bar{y}}_1(a^{n-1}t))\, dt = $$

$$\quad\quad\quad\quad\quad\quad\quad\quad\quad (15.13)$$

$$= a^{m-n+1} \int_0^{t^*a^{n-1}} \phi_0(\bar{\bar{z}}_1(t))\, dt = -\delta_1.$$

By virtue of condition (d) and (15.11) we can estimate the second integral in (15.12):

$$\int_0^t \phi_1(z_1(t))\, dt \leq \int_0^t \beta\left(\frac{M_2}{a}\right) dt \leq T_0\beta(M_2)a^{-m+n-1-q}$$

$$0 \leq t \leq T_0 a^{n-1}. \quad\quad\quad\quad (15.14)$$

Taking (15.11) into account, we can estimate the third integral in (15.12) as follows:

$$\left\|\int_0^t [\phi_0(z_1(t)) - \phi_0(\bar{\bar{z}}_1(t))]\, dt\right\| = $$

$$a^{-m} \left\| \int_0^t [\phi_0(ax_1, y_1) - \phi_0(a\bar{\bar{x}}_1, \bar{\bar{y}}_1)] \, dt \right\| \le$$
$$\le Na^{-m}(\| ax_1 - a\bar{\bar{x}}_1 \| + \| y_1 - \bar{\bar{y}}_1 \|)a^{n-1}T_0 \le \quad (15.15)$$
$$\le 2N\tilde{M}_2 T_0 a^{-m+n-1+k} = M_3 a^{-m+n-1+k}.$$

At the moment $a^{n-1}t^*$, by virtue of (15.12), v is estimated as

$$v \le v_0 + \mu a^{-m+n-1}(M_3 a^{-k} - \delta_1 + T_0\beta(M_2)a^{-q}).$$

Now we fix $\epsilon > 0$ and choose $a^* = a^*(\epsilon, \delta_1, T_0)$ as to ensure the inequalities

$$M_2(a^*)^{-1} < \epsilon_0,$$
$$M_3(a^*)^{-k} \le \frac{\delta_1}{3}, \quad (15.16)$$
$$T_0\beta(M_2)(a^*)^{-q} \le \frac{\delta_1}{3}.$$

This can be done if $k > 0$, i.e. $s > n$, since a was related to v_0 only by the condition $\| x_0 \| = 1$.

Let us choose a number v_1 ($v_1 \le v_0$) such that the moving surfaces $\{v(x, y) = v_1\}$ lie in the $(a^*)^{-1}$-neighbourhood in the coordinates x. We choose $\eta > 0$ such that the surface $\| x \| = \eta$ lies inside the surface $\{v = v_1\}$. Let us consider a solution $z = \bar{z}(t)$ of the system (15.1); this solution starts from a point lying in the η-neighbourhood and at the moment $t = t_1$ reaches the surface $\{v(x, y) = v_1\}$ at a point $\bar{x}_1$. Introducing the notation $b = \| \bar{x}_1 \|^{-1}$, we see that the estimates (15.6) – (15.16) hold for $\bar{z}(t)$ if a is replaced by b. The inequality (15.12) at the moment $b^{n-1}t^*$ takes the form

$$v < v_1 - \mu b^{-m+n-1}\frac{\delta_1}{3} < v_1, \quad (15.17)$$

which means that the solution has returned inside the surface $\{v(x, y) = v_1\}$ after time not exceeding $T_0 b^{-n+1}$ and, moreover, the solution did not leave the ϵ-neighbourhood. Since all the estimates are uniform with respect to the initial point on the surface $\{v = v_1\}$, the system (15.1) is stable in x for $s > n$.

In the case $s = n$, the number $a^*(\epsilon, \delta_1, T_0)$ must be chosen in somewhat different way:

$$2NT_0^2(e^{3NT_0} - 1)(a^*)^{-p} \sum_{i=1}^{2} L_i(M_1) \leq \frac{\delta_1}{6},$$

$$\frac{M_2}{a^*} \leq \epsilon, \tag{15.18}$$

$$T_0\beta(M_2)(a^*)^{-q} \leq \frac{\delta_1}{3}.$$

The chosen parameter μ_0 must obey the conditions

$$\mu_0 \leq 1, \quad \mu_0 2NT_0^2(e^{3NT_0} - 1)\sum_{i=1}^{2} r_i(M_1) \leq \frac{\delta_1}{6}. \tag{15.19}$$

The proof of asymptotic stability for $s \geq n$ goes as follows. Let us assume the contrary, namely, let the solution be arbitrary close to some equipotential surface, but not contained inside it. This is impossible since any solution, having approached an equipotential surface will enter it after finite time.

This contradiction proves asymptotic stability. If the value of (15.15) has the opposite sign, the following theorem of instability is valid.

Theorem 2.17 *Suppose that conditions (a) – (f) are satisfied and $s < n$.*

Then for any $\epsilon > 0$, there exist $\mu_0 > 0$ and $\eta > 0$ such that if $\| x_0 \| < \eta$ and $\mu < \mu_0$, then $\| x(t) \| \leq \epsilon$ for all $t \geq 0$.

Theorems 2.15 and 2.16 are convenient in applications, since, in order to check condition (f), one must integrate along the solutions whose initial data belong to the unit sphere. The minimal number of such integral curves is related to the accuracy of integration σ, the interval length T_0, and the Lipschitz constant N.

2.16   Theorems on Instability

1. In order to investigate instability of systems of the form (1.1), one uses theorems of Lyapunov ([72], [66], [75]) or Chetayev ([8]). If derivative of the Lyapunov or Chetayev function is equal to zero, we deal with a 'neutral' case, where stability or instability of the equilibrium position depends, as before, on properties of the perturbation μR.

The instability can be also investigated ([62]) by making use of the perturbed Lyapunov or Chetayev function $v = v_0 + u$. However, this requires positiveness of the averaged derivative of the function v, calculated by virtue of the system (1.3).

In theorems which we are going to prove below we assume that the perturbation u of the function v_0 is fixed. We note that it is possible to construct u as a polynomial in μ, to wit

$$u = \sum_{k=1}^{m} \mu^k v_k,$$

where the approximate v_k can be obtained from recurrence equations of the type of (7.3). An example of such a problem will be given at the end of this section.

In the theorems presented below we assume that there exist a function $v_0(t, x, y)$ taking positive values in any arbitrarily small neighbourhood of the point $x = 0$.

Following Chetayev, we will use a notion of a '$v_0 > 0$ region' to denote any region in the neighbourhood $\| x \| < h$ ($h < H$) of the origin of the space of variables x, where the function $v_0(t, x, y)$ takes positive values for $t > 0$ and $y \in D$.

After calculating the derivative of v by virtue of the system (1.1), we use the notation (1.5) – (1.8) in formulating theorems on instability.

A region from the neighbourhood of the origin of the space of variables x, $\| x \| < h$, in which ψ takes positive values for $t > 0$ and $y \in D$ is called *the $\psi > 0$ region*. Analogously, we introduce an intersection of the $v_0 > 0$ and $\psi > 0$ regions, provided that it is not empty, and call it *the* $(v_0, \psi) > 0$ *region*.

Theorems 2.1 – 2.15 provide sufficient conditions for stability of the point $x = 0$ for the system (1.1) with the perturbation μR. The equilibrium position $x = 0$ of the system (1.3) will be called (x, μ) – *stable with respect to the perturbations* μR if for any $\epsilon > 0$ $(\epsilon < H)$ there exist $\eta(\epsilon)$ and $\mu_0(\epsilon)$ such that for all $\mu < \mu_0$ all solutions which at the initial moment satisfy the inequality $\| x_0 \| < \eta$, satisfy the inequality $\| x(t) \| < \epsilon$ for $t > 0$.

If stability, in the sense described above, is not present, we say about *instability* of the point $x = 0$ of the system (1.3) *with respect to the perturbations* μR.

In contrast to the definition of Lyapunov stability, in this case the parameter μ must be small. Such a definition of stability is quite close to the definition of stability under permanently acting perturbations. But stability under permanently acting perturbations is achieved at the expense of properties of the unperturbed system (asymptotic stability): the only requirement is that perturbations are in some sense small. However, we will consider the 'neutral case' where stability is achieved as a result of certain properties of the perturbations, since the unperturbed system is simply stable.

2. Let us formulate and prove the theorem which generalizes the Lyapunov theorem on instability for non – stationary motion [72] (in all variables) in the neutral case. The theorem is proved under stringent conditions which simplify the proof, moreover, the mechanism of growth of the function v_0 along the integral curve of the perturbed system (1.1) is becoming clear.

Theorem 2.18 *Suppose that:*

a) There exists a function $v_0(t, x)$ *bounded in* $\| x \| < h$, $t > 0$ *such that any region* $\| x \| < \eta$, $t > 0$ $(\eta < h)$ *possesses a subregion where* $v_0 > 0$.

b)

$$\frac{\partial v_0}{\partial t} + \frac{\partial v_0}{\partial x} f \geq 0 \quad for \quad \| x \| < h, t > 0.$$

c) $\| R(t, x) \| < R_0, \ |\phi(t, x') - \phi(t, x'')| < N \| x' - x'' \| \ for \ \| x \| < h,$ $t > 0.$

d) Uniformly with respect to t_0 there exists the mean $\psi(t_0, x_0)$.

e) For all values of t_0 and x_0 such that $v_0(t_0, x_0) > \alpha$, with $\alpha > 0$,

$$\psi(t_0, x_0) > \delta,$$

where δ depends on α.

Then the equilibrium position of the system (1.3) is unstable with respect to the perturbations μR.

Proof. Let $\eta > 0$ be an arbitrary small parameter. Let us examine a solution of the system (1.1) $x = x(t)$. The initial data x_0 are chosen as to guarantee the condition $\| x_0 \| < \eta, v_0(t_0, x_0) > \alpha$. According to condition (a), such a choice is possible. According to condition (d), for every $\alpha > 0$ there exists $\delta > 0$ for which the condition

$$\psi(t_0, x_0) > \delta$$

is satisfied.

We will examine behaviour of v_0 along this solution, assuming that the solution does not leave the region $\| x \| < h$. Differentiating v_0 along this solution, we obtain

$$\dot{v}_0 = \frac{\partial v_0}{\partial t} + \frac{\partial v_0}{\partial x} f + \mu \phi(t, x(t)). \tag{16.1}$$

In the region $\| x \| < h, \ t > 0$ condition (b) is fulfilled. Thus, by integrating Eq. (16.1) along the solution $x = x(t)$ and by taking condition (b) into account, we obtain the inequality

$$v_0(t, x(t)) \leq v_0(0, x_0) + \mu \int_0^t \phi(t, x(t)) \, dt. \tag{16.2}$$

Let us split the integral above into integrals over the segments Δt_k, with $t_{k+1} - t_k = l$ (the value of l will be chosen later), and then add and subtract similar integrals calculated along the integral curves $x = \bar{x}_k(t)$ of the system (1.3) with initial data $\bar{x}_k(t_k) = x(t_k) = x_k$:

$$\int_0^t \phi(t, x(t)) \, dt = \sum_{k=0}^m \int_{t_k}^{t_k+l} \phi(t, x(t)) \, dt =$$

$$\tag{16.3}$$

$$= \sum_{k=0}^m \left[\int_{t_k}^{t_k+l} \phi(t, \bar{x}_k(t)) \, dt + [\int_{t_k}^{t_k+l} \phi(t, x(t)) - \phi(t, \bar{x}_k(t))] \, dt \right].$$

We will estimate these integrals along the solutions $\bar{x}_k(t)$, by making use of the fact that ψ_0 is positive. By definition of the mean, there exists a function $k(t)$, $\lim_{t \to \infty} k(t) = 0$ such that

$$(t_{k+1} - t_k)^{-1} \int_{t_k}^{t_k+l} \phi(t, \bar{x}_k(t)) \, dt = [\psi_0(t_k, x_k) + k(\Delta t_k)]. \tag{16.4}$$

We choose $\Delta t_k = l$ large enough that

$$|k(\Delta t_k)| < \frac{\delta}{4}, \tag{16.5}$$

where $\delta > 0$ was chosen on the first interval, where $v_0(0, x_0) > \alpha$. According to Lemma 2.2, we get the following estimate for deviation of the solutions $x(t)$ and $x_k(t)$ on each interval of the length l:

$$\| x(t) - x_k(t) \| < \mu R_0 l e^{Nl}. \tag{16.6}$$

Using condition (c) and the inequality (16.6), one can find μ_0 such that for $\mu < \mu_0$ the inequality

$$|\phi(t, x(t)) - \phi(t, \bar{x}_k(t))| \le \frac{\delta}{4} \tag{16.7}$$

holds, hence,

$$\int_{t_k}^{t_k+l} |\phi(t, x(t)) - \phi(t, \bar{x}_k(t))| \le \frac{\delta}{4} \Delta t_k. \tag{16.8}$$

By virtue of the estimates (16.5), (16.8), and the condition $\psi_0(0, x_0) > \delta$, on the first interval we obtain

$$v_0(t, x(t)) \geq v_0(0, x_0) + \mu l[\delta + k(l) - \frac{\delta}{4}] \geq v_0 + \mu l \frac{\delta}{2}. \qquad (16.9)$$

(If restriction is imposed on the sign of the integral $\int_{t_0}^{t} \phi(t, \bar{x}(t)) \, dt$, then the estimate (16.9) follows directly from (16.8)).

Thus, when t reaches l, the function v_0 increases at least by the value $\frac{1}{2}\mu l \delta$, and therefore all estimates can be applied to the next interval. Therefore, for the function $v(t, x(t))$ defined along the solution $x = x(t)$, at the end of the time interval ml the inequality

$$v_0(t, x(t)) \geq v_0(0, x_0) + \frac{1}{2}\mu m l \delta \qquad (16.10)$$

is satisfied.

If one assumes that $x(t)$ never leaves the region $\| x \| < h$ in which the function v_0 is bounded, then, going with t to infinity in Eq. (16.2) (and thus making the number of integrals m in Eq. (16.10) infinite), we get contradiction, since, according to (16.10), v_0 must also tend to infinity. This means that at the certain moment the solution $x = x(t)$ leaves the region $\| x \| < h$, and since the initial data and the perturbations μR may be arbitrarily small, the equilibrium position of the system (1.3) is unstable with respect to the the perturbations μR.

3. Before formulating the general theorem on stability, let us state and prove the following lemma.

Lemma 2.3 *Let the functions $\phi(t, z)$ and $\psi(t_0, z_0)$ be defined by the relations (1.6) and (1.8), with the limit in (1.8) being uniform in t_0 and z_0. Let there exist a summable function $F(t)$ and a non - decreasing function $\chi_2(\alpha) \lim_{\alpha \to 0} \chi_2(\alpha) = 0$ such that for $\| x \| < h, t > 0, y \in D$ we have*

$$|\phi(t, z') - \phi(t, z'')| \leq \chi_2(\| z' - z'' \|)F(t),$$

$$\int_{t_1}^{t_2} F(t) \, dt \leq F_0(t_2 - t_1),$$

where $[t_1, t_2]$ is any finite interval.

Then, for every $\epsilon > 0$ there exists $\eta(\epsilon)$ such that if $\| z_0' - z_0'' \| < \eta(\epsilon)$, then $|\psi(t_0, z_0') - \psi(t_0, z_0'')| < \epsilon$.

Proof. Let us construct the integral curves of the unperturbed system starting from the points z_0' and z_0'' and let us compare the corresponding means.

$$\int_{t_0}^{t_0+T} \phi(t, \bar{z}'(t)) \, dt = T[\psi(t_0, z_0') + k_1(T)],$$

$$\int_{t_0}^{t_0+T} \phi(t, \bar{z}''(t)) \, dt = T[\psi(t_0, z_0'') - k_2(T)],$$

where $k_1(T)$ and $k_2(T)$ tend to zero as $T \to \infty$ uniformly with respect to t_0 and z_0. Subtracting one integral from the other and dividing by T, we obtain

$$\frac{1}{T} \int_{t_0}^{t_0+T} [\phi(t, \bar{z}'(t)) - \phi(t, \bar{z}''(t))] \, dt =$$

$$[\psi(t_0, z_0') - \psi(t_0, z_0'')] + \sum_{i=1}^{2} k_i(T).$$

First we choose T large enough as to guarantee the inequality

$$|k_1(T)| + |k_2(T)| < \frac{1}{2}\epsilon.$$

Then we fix this T and estimate the difference $\phi(t, \bar{z}'(t)) - \phi(t, \bar{z}''(t))$ by the deviation $\bar{z}' - \bar{z}''$. Applying conditions of the Lemma 2.3 to the solutions of the system (1.3), we can choose η sufficiently small that on a finite fixed time interval of the length T, we have the inequality

$$|\phi(t, \bar{z}'(t)) - \phi(t, \bar{z}''(t))| < \frac{1}{2}\epsilon.$$

Lemma is proved.

4.

Theorem 2.19 *Suppose that:*

a) There exists a function $v_0(t, x, y)$ which is positive and bounded for $v_0 > 0$.

b) For $v_0 > 0$

$$\dot{v}_0 = \frac{\partial v_0}{\partial t} + \frac{\partial v_0}{\partial z} f \geq 0.$$

c) There exists a perturbation $u(t, x, y, \mu)$ of the function $v_0(t, x, y)$ and a non – decreasing function $\chi_1(\mu)$ $\lim_{\mu \to 0} \chi_1(\mu) = 0$ such that for $\| x \| < h$, $t > 0$, $y \in D$ we have $|u| \leq \chi_1(\mu)$.

d) There exist summable functions $M(t)$ and $F(t)$ and constants M_0 and F_0 as well as a non – decreasing function $\chi_2(\alpha)$ $\lim_{\alpha \to 0} \chi_2(\alpha) = 0$ such that for $\| x \| < h$, $t > 0$, $y \in D$ we have

$$|\phi(t, z') - \phi(t, z'')| \leq \chi_2(\| z' - z'' \|)F(t),$$

$$\int_{t_1}^{t_2} F(t)\, dt \leq F_0(t_2 - t_1),$$

$$\| R(t, z) \| < M(t),$$

$$\int_{t_1}^{t_2} M(t)\, dt \leq M_0(t_2 - t_1),$$

where $[t_1, t_2]$ is any finite interval.

e) Uniformly with respect to t_0, x_0 from the region $\| x \| < h$, $t > 0$, $y \in D$ there exists a mean $\psi(t_0, x_0, y_0, \mu)$, positive for $\psi > 0$. In the region $(v_0, \psi) > 0$, in an arbitrarily small neighbourhood of the point $x = 0$ (for $\| x \| < h$, $0 < \eta < h$) there exist a point x_0 and numbers $\alpha > 0$, $\delta > 0$ such that for any $t > 0$, $y \in D$ the inequalities

$$v_0(t_0, x_0, y_0) > 2\alpha,$$

$$\psi(t_0, x_0, y_0) > \delta.$$

f) There exists a number l_1 such that on any segment of the integral curve of the system (1.3) in the domain $\| x \| < h$, $t > 0$, $y \in D$ there exist points from the region $(v_0, \psi) > 0$, for whose the inequality $\psi > \delta$ holds, provided $v_0(t_0, x_0, y_0) > \alpha$.

If all these conditions are satisfied, then the equilibrium position of the system (1.3) is unstable with respect to the perturbation μR.

Proof. Let us choose the initial data x_0 of the system (1.1) to be arbitrarily small, but still satisfying the inequalities $v_0(0, x_0, y_0) > 2\alpha$, $\psi(0, x_0, y_0) > \delta$; according to conditions (a) and (e) of the theorem, this can be done. Consider the perturbed function $v = v_0 + u$ on the solution $x = x(t)$, $y = y(t)$ of the system (1.1), with the initial data x_0, y_0, and assume that this solution never leaves the region $\| x \| < h$. By virtue of the system (1.1), in notation introduced in (1.5) and (1.6), the derivative has the form

$$\dot{v} = \frac{\partial v_0}{\partial t} + \frac{\partial v_0}{\partial z} f + \theta(\mu) \cdot \phi(t, z). \tag{16.11}$$

During the time when the curve $x = x(t)$, $y = y(t)$ remains in the region $v_0 > 0$, condition (b) holds, and therefore, by integrating (16.11), we obtain

$$v_0(t, z(t)) \geq v_0(0, z_0) + \theta(\mu) \int_0^t \phi(t, z(t))\, dt \tag{16.12}$$

or

$$v_0(t, z(t)) \geq v_0(0, z_0) + u(0, z_0, \mu) - \tag{16.13}$$

$$-u(t, z(t), \mu) + \theta(\mu) \int_0^t \phi(t, z(t))\, dt.$$

We choose μ_1 sufficiently small as to guarantee that for $0 < \mu < \mu_1$

$$2|u| < \alpha; \tag{16.14}$$

according to conditions of the theorem, this is possible.

In order to estimate the integral in Eq. (16.12), we add and subtract the integral calculated along the curve $z = \bar{z}(t)$, $\bar{z}(0) = \bar{z}_0$ of the system (1.3)

$$\int_0^t \phi(t, z(t))\, dt = \int_0^t \phi(t, \bar{z}(t))\, dt + \tag{16.15}$$

$$+ \int_0^t [\phi(t, z(t)) - \phi(t, \bar{z}(t))]\, dt.$$

According to definition of the mean, there exists a function $k(t)$ such that $\lim\limits_{t\to\infty} k(t) = 0$ and

$$\int_0^t \phi(t, \bar{z}(t))\, dt = t[\psi(0, z(0)) + k(t)]. \tag{16.16}$$

We choose l sufficiently large as to guarantee the inequality

$$\| k(t) \| < \frac{\delta}{8}. \tag{16.17}$$

Now, let us examine the inequalities (16.12) and (16.13) on the time interval L ($L = l + l_1$), where l_1 is given by condition (f) of the theorem. We choose μ_2 such that $\mu < \mu_2$ on the time interval $[0, L]$ and by virtue of condition (d) and Lemma 2.2, we have the estimate

$$\int_0^t \| \phi(t, z(t)) - \phi(t, \bar{z}(t)) \| \, dt < \frac{\delta t}{8}. \tag{16.18}$$

We impose one more condition on μ, namely that $\mu < \mu_3$, where μ_3 is such that on a time interval of the length L the curve $x = x(t)$, $y = y(t)$ stays within the region $v_0 > 0$. To this end we need the inequality

$$\theta(\mu_3)|\int_0^L \phi(t, z(t))\, dt| < \frac{1}{2}\alpha. \tag{16.19}$$

Here, we assumed that $\theta(\mu)$ is a non – decreasing function, $\lim\limits_{\mu\to 0}\theta(\mu) = 0$. Combining these assumptions, condition (e), and the inequality (16.18), we can choose μ_3 such that the inequality (16.19) is satisfied. Thus, for $\mu < \mu_3$, the curve $x = x(t)$, $y = y(t)$ stays in the region $v_0 > 0$ for $0 < t \leq L$ and the inequality (16.12) is satisfied.

According to condition (f) of the theorem, there exist a point $(t_1, \bar{z}(t_1))$ belonging to the time interval $[l, l + l_1]$, where $\psi(t_1, \bar{z}(t_1)) > \delta$ We impose one more restriction on the parameter μ, namely $\mu < \mu_1$, then at some moment t_1 the mean $\psi(t_1, z(t_1))$ satisfies the condition

$$\psi(t_1, z(t_1)) > \frac{1}{2}\delta. \tag{16.20}$$

on the curve $z = z(t)$ as well.

To do that, with the help of Lemma 2.3 we choose the deviation of the points z_1 and $\bar{z}_1$ so small that the difference $\psi(t_1, z(t_1)) - \psi(t_1, \bar{z}(t_1))$ does not exceed $\delta/4$. Further, using Lemma 2.3, we choose μ_4 small enough as to make the solutions of the systems (1.1) and (1.3) diverging by a sufficiently small distance.

Now we introduce $\mu_0 = \min\{\mu_1, \ldots, \mu_4\}$ and consider the inequality (16.12) for $0 < \mu < \mu_0$; this inequality can be used to estimate $v_0(t_1, z(t_1))$.

To do this, we combine condition (e), the inequalities (16.12) to (16.14), relation (16.15), and inequalities (16.17), (16.18). As a result, we obtain the estimate

$$v_0(t_1, z(t_1)) \geq \alpha + \theta(\mu)[\psi(0, z(0)) - \frac{\delta}{4}] \geq$$

$$(16.21)$$

$$\geq \alpha + \theta(\mu)l(\delta - \frac{\delta}{4}).$$

Now we extend the integral curve of the system (1.1): from the point (t_1, z_1) to the next point (t_2, z_2), which we choose in exactly the same way as we did for (t_1, z_1); by exactly the same arguments the inequalities (16.17) – (16.19) will be fulfilled. For $v_0(t_2, z(t_2))$ we obtain the estimate

$$v_0(t_2, z(t_2)) \geq \alpha + l\theta(\mu)\frac{3}{4}\delta + \lambda\theta(\mu)\left[\frac{\delta}{2} - \frac{\delta}{4}\right]. \qquad (16.22)$$

By extending the integral curve $z = z(t)$ still farther, we obtain the estimate for $v_0(t_m, z(t_m))$, to wit

$$v_0(t_m, z(t_m)) \geq \alpha + l\theta(\mu)\frac{3}{4}\delta + (m - 1)\lambda\theta(\mu)\frac{\delta}{4}, \qquad (16.23)$$

from which we see that $v_0(t, z(t)) \to \infty$ as $t \to \infty$, which contradicts the assumption that v_0 is bounded for $\| x \| < h$, $t > 0$, $y \in D$. Thus, the integral curve does leave the region $\| x \| < h$ and the equilibrium position of the system (1.3) is unstable with respect to the perturbations μR. The theorem is proved.

We should note that the theorem proved above is related to the case when the perturbation u of the Lyapunov function is given not in an annular region, but in the entire neighbourhood of the equilibrium position. In terms of the notion of instability, the technique of function u defined in an annular region is not appropriate to define instability.

Also note that if the mean $\psi(t_0, x_0, y_0, \mu)$ does not exist, then the restriction can be imposed directly on the sign of the integral (16.16).

In some cases condition (f) may be replaced by the requirement that solutions of the system (1.3) are periodic.

5. As an illustrative example we examine the Mathieu equation with small values of the parameter h and with periodic coefficients

$$\ddot{x} + \omega^2(1 - h\cos \nu t)x = 0. \tag{16.24}$$

Let us rewrite this equation as the system

$$\dot{x} = y, \quad \dot{y} = -\omega^2 x + \omega^2 h x \cos \nu t. \tag{16.25}$$

The term $\omega^2 h x \cos \nu t$ where h is small is regarded as a perturbation.

The unperturbed system has the solution

$$\bar{x} = x_0 \cos \omega(t - t_0) + \frac{y_0}{\omega} \sin \omega(t - t_0),$$
$$\bar{y} = -x_0 \omega \sin \omega(t - t_0) + y_0 \cos \omega(t - t_0), \tag{16.26}$$
$$\bar{x}(t_0) = x_0, \quad \bar{y}(t_0) = y_0.$$

The function $v = \omega^2 x^2 + y^2$ is the first integral of the unperturbed system. The time derivative of this function is by virtue of (4.25)

$$\dot{v} = 2\omega^2 h x y \cos \nu t = h\phi(x, y, t). \tag{16.27}$$

Let us calculate the mean along the trajectories of (16.26) with $\nu = 2\omega$:

$$\psi(t_0, x_0, y_0) = \frac{x_0^2 \omega^2 - y_0^2}{2} \sin 2\omega t_0 + x_0 y_0 \omega^2 \cos 2\omega t_0. \tag{16.28}$$

Being a mean, the function $\psi(t, x, y)$ is a first integral of the system (16.26). We regard it as a Chetayev – type functional. It changes the sign in the neighbourhood of 0 (positive and negative sectors rotate around the origin with the angular velocity $\dfrac{2\pi}{2\omega}$),

$$\dot{\psi} = h[\omega^4 x^2 \cos^2 2\omega t - \omega^2 xy \sin 2\omega t \cos 2\omega t] = h\phi_1(t, x, y).$$

Let us now calculate ψ_1 which is the mean of ϕ_1

$$\psi_1(x, y) = \omega^4 \frac{\omega^2 x^2 + y^2}{4\omega^2}$$

and is positive. According to Theorem 2.18, the equilibrium position $x,, y$ is h – unstable. In order to find other resonances, one should construct perturbation of the Lyapunov function.

2.17   Study of Stability of Perturbed Systems Using Positive – Definite Function which is not Lyapunov Function

1. Consider a perturbed system

$$\dot{x} = f(t, x) + \mu R(t, x), \tag{17.1}$$

with small parameter μ, $0 < \mu < 1$, in the region $G = I \times B_H$, $I = \{t \in \mathbf{R} | t \geq 0\}$, $B_H = \{x \in \mathbf{R}^n | \parallel x \parallel < H\}$, assuming that the unperturbed system

$$\dot{x} = f(t, x), \quad f(t, 0) = 0 \tag{17.2}$$

has a stable equilibrium point $x = 0$.

In the previous sections of this chapter, we suggested a generalization of Lyapunov second method based on the assumption that the Lyapunov function of the unperturbed system is known and that the conditions of the Lyapunov theorem on stability are satisfied. Here we consider a more general situation when the equilibrium position $x = 0$

of the system (17.2) is stable and the function $v(t, x)$ is known. This function is positive – definite, admits an infinitely small upper bound, and satisfies one of the theorems of the comparison method ([19], [85], [98]): its derivative satisfies the inequality

$$\frac{dv}{dt} = \frac{\partial v}{\partial t} + \nabla v \cdot f(t, x) \leq \omega(t, v(t, x)). \tag{17.3}$$

The function $\omega(t, u)$ is such that the scalar equation of comparison

$$\dot{u} = \omega(t, u) \tag{17.4}$$

has a solution stable with respect to positive initial perturbations. Thus, the function is not required to be of the Lyapunov type.

By $u_0(t, t_0, \Delta)$ we denote the solution of the Cauchy problem

$$\dot{u}_0 = \omega(t, u_0), \quad u_0(t_0) = \Delta. \tag{17.5}$$

Let $\Phi(t, t_0)$ denote the solution of the linearized problem

$$\dot{\bar{u}} = \frac{\partial \omega}{\partial u}(t, u_0(t, t_0, \Delta))\bar{u}, \quad \bar{u}(t_0) = 1,$$

$$\Phi(t, t_0) = \exp(\int_{t_0}^{t} \frac{\partial \omega}{\partial u}(\tau, u_0(\tau, t_0, \Delta))\bar{u}\, d\tau). \tag{17.6}$$

Let us also introduce the function

$$\phi(t, x) = \nabla v \cdot R(t, x).$$

2.

Theorem 2.20 *([3]) Suppose that in the region* $\| x \| \leq H$, $t \geq 0$:

a) There exist a continuously differentiable function $v(t, x)$ *and continuous functions* $a(\alpha)$, $b(\alpha)$ *($a(0) = b(0) = 0$) monotonically increasing for* $\alpha \geq 0$ *such that*

$$a(\| x \|) \leq v(t, x) \leq b(\| x \|).$$

b) The function $v(t, x)$ satisfies the inequality (17.3), where the function $\omega(t, u)$ is continuously differentiable with respect to u in the region $\{t \geq 0, u \geq 0\}$.

c) There exist summable functions $M(t)$, $L(t)$ and constants M_0, L_0, as well as a function $\sigma(\alpha)$ $(\sigma(0) = 0)$ monotonically increasing for $\alpha \geq 0$ such that

$$|\phi(t, x') - \phi(t, x'')| \leq L(t)\sigma(\| x' - x'' \|),$$

$$\| R(t, x) \| \leq M(t),$$

$$\int_{t_1}^{t_2} M(t)\, dt \leq M_0(t_2 - t_1),$$

$$\int_{t_1}^{t_2} L(t)\, dt \leq L_0(t_2 - t_1)$$

for any finite interval $[t_1, t_2]$.

d) For any sufficiently small $\eta > 0$ there exist $T(\eta) \geq 0$ and $\delta = \delta(\eta)$ such that

$$\int_{t_0}^{t_0+T} \phi(t, \bar{x}(t, t_0, x_0))\Phi(t_0 + T, t)\, dt \leq -\delta T(\eta) < 0$$

for any $t_0 \geq 0$ and $\eta \leq \| x_0 \| \leq H$.

Then for every $\epsilon > 0$ and $\tau_0 \geq 0$ there exist $\eta = \eta(\epsilon) > 0$ and $\mu_0 = \mu_0(\epsilon) > 0$ such that for $0 < \mu \leq \mu_0$ and $\| x(\tau_0, \mu) \| \leq \eta$ the solution $x = x(t, \mu)$ of the system (17.1) is defined for all $t \geq \tau_0$ and $\| x(t, \mu) \| \leq \epsilon$ for all $t \geq \tau_0$.

Proof. Let us fix arbitrary $\epsilon > 0$ $(\epsilon < H)$. Let $\Delta = \Delta(\epsilon) > 0$ be such that $u_0(t, t_0, u^0) \leq \frac{1}{2}a(\epsilon)$ for $t \geq t_0$, if $0 < u^0 \leq \Delta(\epsilon)$. By S_Δ we denote the moving surface $v(t, x) = \Delta$. By virtue of condition (a) of the theorem $\| x \| \geq b^{-1}(\Delta) = \eta > 0$ for $x \in S_\Delta$. Let $x(t, \mu)$ be a solution of Eq. (17.1) whose trajectory begins in the η-neighbourhood of the point $x = 0$. Let $x_0 = x(t_0, \mu) \in S_\Delta$ at some $t_0 > 0$, i.e. $v(t_0, x_0) = \Delta$. To estimate the variation of $v(t, x(t, \mu))$ for $t \geq t_0$, let us consider the expression for the total derivative of the function $v(t, x)$ by virtue of the system (17.1). Taking condition (b) into account, we obtain

$$\frac{dv}{dt} = \frac{\partial v}{\partial t} + (\nabla v)f(t, x) + \mu\phi(t, x) \leq \omega(t, v(t, x)) + \mu\phi(t, x). \quad (17.7)$$

Let us examine a scalar equation with the initial data

$$\dot{u} = \omega(t, u) + \mu\phi(t, x(t, \mu)), \quad u(t_0, \mu) = \Delta. \quad (17.8)$$

We represent the solution $u(t, \mu)$ of this problem as

$$u(t, \mu) = u_0(t) + \mu\tilde{u}(t, \mu), \quad (17.9)$$

where $u_0(t)$ is the solution of (17.5). Then, for $\tilde{u}(t, \mu)$ we have

$$\dot{\tilde{u}} = \frac{\partial\omega}{\partial u}(t, u_0(t))\tilde{u} + \phi(t, x(t, \mu)) + \Omega(t, \mu, \tilde{u}),$$
$$(17.10)$$
$$\tilde{u}(t_0, \mu) = 0.$$

For bounded $|\tilde{u}|$, uniformly with respect to $t \geq 0$, we have $\Omega(t, \mu, \tilde{u}) \rightarrow 0$ as $\mu \rightarrow 0$. We now consider the Cauchy problem for the non – homogeneous equation.

$$\dot{\tilde{u}} = \frac{\partial\omega}{\partial u}(t, u_0(t))\tilde{u} + \phi(t, x(t, \mu)),$$
$$(17.11)$$
$$\bar{u}(t_0) = 0,$$

which is obtained from Eq. (17.10) by omitting nonlinear terms on its RHS. If Φ is the solution of Eq. (17.5) satisfying $\Phi(\tau, \tau) = 1$, then the function

$$\bar{u}(t, t_0) = \int_{t_0}^{t} \phi(\tau, x(\tau, \mu))\Phi(t, \tau) \, d\tau \quad (17.12)$$

is the solution of problem (17.11).

Taking into consideration the fact that the parameter μ is small, it is not difficult to show that there exists $\mu_0 > 0$ such that for $0 < \mu \leq \mu_0$, condition (d) of the theorem implies

$$\bar{u}(t_0 + T, t_0) \leq -\frac{3}{4}\delta < 0. \tag{17.13}$$

The continuous function $\bar{u}(t, t_0)$ is uniformly bounded with respect to $t_0 \geq 0$ on the interval $t_0 \leq t \leq t_0 + T$:

$$\bar{u}(t, t_0) \leq U_0, \quad U_0 = \text{const} > 0.$$

Assuming that $|\tilde{u}| \leq U_0$ for $t \in [t_0, t_0 + T]$ and taking Eqs. (17.10) and (17.11) into account, we see that for a function $\chi(a)$ ($\chi(0) = 0$) monotonically increasing for $t \in [t_0, t_0 + T]$

$$|\tilde{u}(t, \mu) - \bar{u}(t)| \leq \chi(\mu). \tag{17.14}$$

Let $\chi(\mu) \leq \frac{1}{4}\delta$ for $0 < \mu \leq \mu_0$. Then

$$\tilde{u}(t_0 + T), \mu) \leq \bar{u}(t_0 + T) + \chi(\mu) \leq -\frac{1}{2}\delta < 0, \tag{17.15}$$

where $\tilde{u}(t, \mu) \leq U_0$ for $t \in [t_0, t_0 + T]$. Thus,

$$u(t, \mu) = u_0(t) + \mu\tilde{u}(t, \mu) \leq a(\epsilon) \tag{17.16}$$

for $t \in [t_0, t_0 + T]$, provided that $0 < \mu \leq \mu_0$ and that μ_0 is sufficiently small that $\mu_0 U_0 \leq \frac{1}{2}a(\epsilon)$. If, for some value $t_1 > t_0 + T$, the equation $\tilde{u}(t_1, \mu) = 0$ holds again, then, given the uniformity of the estimates with respect to t_0, we have $\tilde{u}(t_1 + T, \mu) \leq -\frac{1}{2}\delta$ and the inequality (17.16) remains valid in the interval $[t_1, t_1 + T]$ as well.

Thus, the inequality (17.15) remains valid for all $t < t_0$, for which $\| x(t, \mu) \|$ exceeds $\eta = \eta(\epsilon) > 0$. If for some τ_1 we have $\| x(\tau_1, \mu) \| < \eta$, this means that the trajectory $x(t, \mu)$ has returned inside the surface S_Δ. The trajectory $x(t, \mu)$ may leave the region $\| x \| < \eta$ many times, while remaining inside the region $\| x \| < \epsilon$, since according to the Chaplygin lemma on comparison, we have

$$a(\| x(t,\mu)) \|) \leq v(t, x(t,\mu)) \leq u(t,\mu) \leq a(\epsilon)$$

for $\| x(t,\mu) \| \geq \eta$. The theorem is proved.

3. Let us now consider the system (17.1), assuming that the so – called Lyapunov vector function ([85], [73]) is known for the unperturbed system (17.2).

Let D be some neighbourhood of zero in $\mathbf{R}^m$. We introduce the comparison system

$$\dot{u} = F(t,u), \quad F(t,0) = 0, \qquad (17.17)$$

where $F : I \times D \to \mathbf{R}^m$ is differentiable twice with respect to u and increasing quasi – monotonically in the sense of the definition given in Ref. [98]. The inequality $u \geq z$ for arbitrary $u, z \in \mathbf{R}^m$ means, as in [98] that $u_i \geq z_i$, $i = \overline{1,m}$. The inequality $u > z$ is defined in a similar fashion. We denote the solution of the system (17.17) with the initial condition $u(t_0; t_0, \Delta) = \Delta$ by $u(t; t_0, \Delta)$. Let $\bar{e}$ be a vector in $\mathbf{R}^m$, all components of which are equal to 1. In addition, as in [98], we denote by K the class $\{a(\alpha)\}$ of continuous functions monotonically increasing for $\alpha \geq 0$, $a(0) = 0$. It is known, [98] that existence of the function $v(t,x) = \{v_1(t,x), v_2(t,x), \ldots, v_m(t,x)\}$ which satisfies the conditions

$$a(\| x \|) \leq \max_{1 \leq i \leq m} v_i(t,x) \leq b \| x \|, \qquad (17.18)$$

$$\frac{\partial v}{\partial t} + \frac{\partial v}{\partial x} f(t,x) \leq F(t,v) \qquad (17.19)$$

for some $a, b \in K$ implies the uniform stability of the zero solution of the unperturbed system (17.2), if the zero solution of the unperturbed comparison system is stable in the following sense: $(\forall \epsilon > 0) \, (\exists \gamma > 0) \, (\forall t_0 \in I) \, (\forall \Delta \leq \gamma \bar{e}) \, (\forall t \geq t_0) \, u(t; t_0, \Delta) \leq \epsilon \bar{e}$.

Let us use the following notation $\phi_i(t, x) = (\nabla v_i)R(t, x)$, $i = \overline{1, m}$. $\phi = \{\phi_1, \ldots, \phi_m\}$. Thus, $\phi(t, x) = \dfrac{\partial v}{\partial x} R(t, x)$. Let $\Phi(t, s, t_0, \Delta)$ be a matrixant of the system of variations

$$\dot{r} = \frac{\partial F}{\partial u}(t, u(t, t_0, \Delta))r \qquad (17.20)$$

i.e. $\Phi(t, t, t_0, \Delta) = E$ is the unit matrix.

4.

Theorem 2.21 *([4]). Suppose that in the domain G:*

a) There exists a continuously differentiable vector – valued function $v(t, x) = \{v_1(t, x), v_2(t, x), \ldots, v_m(t, x)\}$ and a function $F(t, u)$, with both functions satisfying conditions presented above.

b) There exist summable functions $\mathcal{M}(t)$, $\mathcal{L}(t)$ and constants $\mathcal{M}_0$, $\mathcal{L}_0$, as well as a function $\sigma \in K$ such that

$$|\phi(t, x') - \phi(t, x'')| \leq \mathcal{L}(t)\sigma(\| x' - x'' \|),$$

$$\| R(t, x) \| \leq \mathcal{M}(t),$$

$$\int_{t_1}^{t_2} \mathcal{M}(t)\, dt \leq \mathcal{M}_0(t_2 - t_1),$$

$$\int_{t_1}^{t_2} \mathcal{L}(t)\, dt \leq \mathcal{L}_0(t_2 - t_1)$$

for any finite interval $[t_1, t_2] \subset I$.

d) For any sufficiently small $\eta > 0$ there exist $T = T(\eta) > 0$ and $\delta = \delta(\eta)$ such that

$$\int_{t_0}^{t_0+T} \Phi(t_0 + T, t, t_0, \Delta)\phi(t, \bar{x}(t, t_0, x_0))\, dt \leq -\delta\bar{e},$$

where $\Delta = v(t_0, x_0)$ for any $t_0 > 0$ and $x_0 \in B_H/B_\eta$.

Then for every $\epsilon > 0$ there exist $\eta = \eta(\epsilon) > 0$ and $\mu_0 = \mu_0(\epsilon) > 0$ such that for any μ, $0 < \mu \leq \mu_0$, $\tau_0 \geq 0$, and $x_0 \in B_\eta$ the solution $x = x(t; \tau_0, x_0)$ of the system (17.1) is defined for all $t \geq \tau_0$ with $\| x(t; \tau_0, x_0) \| \leq \epsilon$.

Proof. Let us denote $w(x,t) = \max\limits_{1 \leq j \leq m} v_j(t,x)$. We fix an arbitrarily small number $\epsilon > 0$. Let $\gamma = \gamma(\epsilon) > 0$ be such that

$$1)\ 0 < \gamma < a^{-1}(\epsilon); \quad 2)\ u(t; t_0, \Delta) \leq \frac{1}{2}a(\epsilon)\bar{e}, \quad \Delta \leq \gamma\bar{e} \qquad (17.21)$$

where $a(\alpha) \in K$ and satisfies (17.4). By $\rho_\gamma = \{x \in B_h | w(t,x) = \gamma\}$ we denote a moving surface which includes the neighbourhood of zero $G_\gamma(t)$ in $\mathbf{R}^n$. Conditions (17.18) and (17.21) imply that

$$B_\eta \subset G_\gamma(t) \subset B_\epsilon \qquad (17.22)$$

for all $t \geq 0$, if $\eta \leq b^{-1}(\gamma)$. A trajectory of the solution $x(t)$ of the system (17.1) starts in the ball B_η and then crosses the surface $S_\gamma(t_0)$ at the point x_0, i.e. $w(t_0, x_0) = \gamma$,

$$v(t_0, x_0) \leq \gamma\bar{e}. \qquad (17.23)$$

Let us estimate the change of the vector – valued function $v(t, x(t))$ for $t > t_0$, by taking the inequality (17.19) into account:

$$\dot{v}|_1 = \dot{v}|_2 + \mu\phi(t, x(t)) \leq F(t, v) + \mu\phi(t, x(t)). \qquad (17.24)$$

Let us consider the auxiliary Cauchy problem

$$\dot{U} = F(t, U) + \mu\phi(t, x(t)), \quad U(t_0, \mu) = U(t_0, x_0). \qquad (17.25)$$

We represent its solution as

$$U(t, x) = u(t) + \mu\tilde{U}(t, \mu), \qquad (17.26)$$

where $u(t) = u(t, t_0)$ and $U(t_0, \mu)$ a solution of the comparison system (17.17). For the function $\tilde{u}$ we have the problem

$$\dot{\tilde{u}} = \frac{\partial F}{\partial u}(t, u(t))\tilde{u} + \phi(t, x(t)) + \Omega(t, \tilde{u}, \mu),$$
$$\tilde{u}(t_0, \mu) = 0. \qquad (17.27)$$

Because of the properties of the function $F(t, u)$, we have $F(t, u) \to 0$ as $u \to 0$ uniformly with respect to $t \geq 0$ and $\| \tilde{u} \| \leq$ const. Let $\bar{u}(t)$ be a solution of the problem

$$\dot{\bar{u}} = \frac{\partial F}{\partial u}(t, u(t))\bar{u} + \phi(t, x(t)),$$
$$\bar{u}(t_0) = 0. \qquad (17.28)$$

Then

$$\bar{u}(t) = \int_{t_0}^{t} \Phi(t, \tau)\phi(\tau, x(\tau))\, d\tau,$$

where $\Phi(t, \tau)$ is the matrixant of the linear system (17.20) with $\Delta = v(t_0, x_0)$. If $\bar{x}(t)$ is the solution of the unperturbed system (17.2) with the initial data $\bar{x}(t) = x(t_0) = x_0$, then the interval $t_0 \leq t \leq t_0 + T$ (with $T = T(\eta)$ being determined by condition (c)), $\| x(t) - \bar{x}(t) \| < \mu$const.

Thus,

$$\bar{u}(t_0 + T) + \int_{t_0}^{t_0+T} \Phi(t_0 + T, \tau)\phi(\tau, x(\tau))\, d\tau \leq -\frac{3}{4}\delta\bar{e}.$$

Hence, for sufficiently small μ, comparing the solutions of the problems (17.27) and (17.28), we obtain

$$\tilde{u}(t_0 + T, \mu) \leq -\frac{\delta}{2}\bar{e}. \qquad (17.29)$$

By virtue of (17.21), (17.23), and (17.29), the representation (17.26) implies that on the interval $t_0 \leq t \leq t_0 + T$, for any sufficiently small μ

$$U(t, \mu) \leq a(\epsilon)\bar{e}. \qquad (17.30)$$

Moreover,

$$U(t_0 + T, \mu) < u(t_0 + T) \leq \frac{1}{2}a(\epsilon)\bar{e}. \qquad (17.31)$$

Using the comparison lemma of ref [98] and (17.17), from (17.30) we obtain the inequality $\| x(t) \| < \epsilon$ on the interval $t_0 \leq t \leq t_0 + T$.

Let us denote $t_1 = t_0 + T$, $x_1 = x(t_1)$. If $w(t_1, x_1) \leq \gamma$, then we know that the trajectory $x(t)$ returned to the domain $G_\gamma(t)$. By virtue of (17.22), the inequality $\| x(t) \| < \epsilon$ holds while $x(t) \in G_\gamma(t)$. If the trajectory leaves $x(t)\, G_\gamma(t)$ again, then, as a result of the uniformity of the above estimates with respect to $t_0 \geq 0$ and x_0, $\| x_0 \| \geq \eta$, we can repeat our arguments and obtain the inequalities (17.30) and (17.31) again. Let now $w(t_1, x_1) > \gamma$. Note that then $\| x_1 \| > \eta$. Combining (17.24), (17.25), (17.31), and the comparison lemma we get

$$u(t_1, x_1) \leq U(t_1, \mu) < u(t_0 + T) = u(t_0 + T; t_0, v(t_0, x_0)).$$

Again, by virtue of the comparison lemma for all $t > t_1$

$$u_1(t) = u(t, t_1, v(t_1, x_1)) < u(t) = u(t, t_0, v(t_0, x_0)).$$

Hence, $u_1(t) \leq \frac{1}{2}a(\epsilon)\bar{e}$ for $t \geq t_1$. Denoting by $U_1(t, \mu)$ the solution of the system (17.25) with the initial data $U_1(t_1, \mu) = v(t_1, x_1)$ and repeating the arguments presented above, we get the inequality $U_1(t, \mu) \leq a(\epsilon)\bar{e}$ for $t_1 \leq t \leq t_1 + T$. Then it follows that the inclusion $x(t) \in B_\epsilon$ is valid for $t \in [t_1, t_1 + T]$ as well. Finally, let us denote $t_2 = t_1 + T$, $x_2 = x(t_2)$. Again we have that either $w(t_2, x_2) \leq \gamma$ or $w(t_2, x_2) > \gamma$. Since the estimates are uniform with respect to $t_0 \geq 0$ and $x_0 \in B_H/B_\eta$, the situation described above happens again and again infinite number of times, and always $x(t) \in B_\epsilon$. The theorem is proved.

5. The fact that the integral

$$J(t_0, x_0, T) =$$
$$\int_{t_0}^{t_0+T} \Phi(t_0 + T, t, t_0, v(t_0, x_0))\phi(t, \bar{x}(t; t_0, x_0))\, dt$$

is positive under certain assumptions imposed on $v(t, x)$ is a sufficient condition of instability of the system (17.1). Let us start by assuming that the system (17.17) is linear

$$\dot{u} = F(t, u) = A(t)u. \tag{17.32}$$

In this case, the system of variations (17.20) coincides with the comparison system itself, and therefore, its matrixant Φ does not depend on t_0 and Δ.

Let us introduce a notion of the so – called positiveness region of the vector function $v(t, x)$: $\{v > 0\} = \{(t, x) : v_1(t, x) > 0, \ldots, v_m(t, x) > 0\}$. The region $\{v > 0\}$ is said to adjoin the origin if $v(t, 0) = 0$ and if, for arbitrarily small $\eta > 0$, there exists x_0, $\| x_0 \| < \eta$ such that $v(t_0, x_0) > 0$, i.e. $(t_0, x_0) \in \{v > 0\}$.

Theorem 2.22 *Suppose that in the region G:*

a) There exists a continuously differentiable vector – valued function $v(t, x) = \{v_1(t, x), v_2(t, x), \ldots, v_m(t, x)\}$ whose positiveness region adjoins the origin.

b) The solution $u(t; t_0, u_0)$ of the comparison system possesses the following property: $\forall u_0$, $u(t; t_0, u_0) \geq \rho(t_0, u_0)$ for $t \geq t_0$.

c) In the region $\{v > 0\}$, $\dot{v}|_2 \geq F(t, v)$.

d) There exists a function $b \in K$ such that for at least one component $v_k(t, x)$ of the vector – valued function v the estimate $v_k(t, x) \leq b(\| x \|)$ is valid in the region $\{v > 0\}$.

e) For any small α there exist $T = T(\alpha) > 0$ and $\delta = \delta(\alpha)$ such that $J(t_0, x_0, T) \geq \delta \bar{e}$ if $v(t_0, x_0) \geq \alpha \bar{e}$.

f) Condition (b) of Theorem 2.20 is satisfied.

Then, there exists $h > 0$ such that for any $\eta > 0$ and $\mu_0 > 0$ there exist $t_0 \geq 0$, $x_0 \in B_\eta$, μ, $t_1 > t_0$ such that $\| x(t_1, t_0, x_0) \| > h$.

Proof. The proof is by contradiction. Let us assume that there exists small $\eta > 0$ and $\mu_0 > 0$ such that for all x_0, $\| x_0 \| < \eta$ we have $x(t) \in B_h$ for some $h > 0$, all $t > t_0$, and $0 < \mu < \mu_0$. We choose a point $x_0 \in B_\eta$ such that $v(t_0, x_0) > 0$. By virtue of condition (b) of the theorem, there exists $\alpha > 0$ such that the solution of the comparison system (17.32) $u(t) = u(t; t_0, v(t_0, x_0))$ satisfies the inequality

$$u(t) \geq \alpha \bar{e} \qquad (17.33)$$

for all $t \geq t_0$. Let us estimate how $v(t, x(t))$ changes along the solution $x = x(t; t_0, x_0)$ of the system (17.1). Assuming that $(t, x(t)) \in \{v > 0\}$, we obtain

$$\dot{v}|_1 \geq A(t)v(t, x(t)) + \mu\phi(t, x(t)). \qquad (17.34)$$

Consider the auxiliary Cauchy problem

$$\dot{U} = A(t)U + \mu\phi(t, x(t)), \quad U(t_0, \mu) = v(t_0, x_0). \qquad (17.35)$$

Knowing the matrixant of the system (17.32), we can write down the solution of the system (17.35) explicitly, to wit

$$U(t, \mu) = \Phi(t, t_0)v(t_0, x_0) + \mu \int_{t_0}^{t} \Phi(t, \tau)\phi(\tau, x(\tau))\, d\tau.$$

We denote $t_i = t_0 + iT$, where the constant $T(\alpha) > 0$ is determined by condition (e). Then, for any integer $s > 0$, we get

$$U(t_s, \mu) = \Phi(t_s, t_0)v(t_0, x_0) + \mu \sum_{i=0}^{s-1} \int_{t_i}^{t_{i+1}} \Phi(t_{i+1}, \tau)\phi(\tau, x(\tau))\, d\tau.$$

By virtue of (17.33) and condition (e), for sufficiently small $\mu_0 > 0$ and all μ, $0 < \mu < \mu_0$, we have $U(t_s, \mu) \geq \alpha\bar{e} + \frac{1}{2}\delta s\bar{e}$. By virtue of the comparison lemma, $v(t_s, x_s) \geq U(t_s, \mu)$, where, as one can see, $U(t_s, \mu) \to \infty$ as $s \to \infty$. By condition (d), for some k, $1 \leq k \leq m$, $b(\|\, x\, \|) > v_k(t, x)$, and therefore, $\|\, x(t_{s-1})\, \| \to \infty$ as $s \to \infty$, and this contradicts the assumption that $\|\, x(t)\, \| < h$. The theorem is proved.

6. Let us now turn to the general case of nonlinear comparison system. Sufficient conditions of instability are established by the following theorem.

Theorem 2.23 *Suppose that in the domain G:*

a) Conditions (a) and (c) to (f) of Theorem 2.21 are satisfied.

b) The solution $u(t; t_0, u_0)$ of the comparison system has the following property: $\forall u_0 > 0$, $\forall t_0 > 0$, and for any small $\delta_1 > 0$ the following *inequalities hold*

$$u(t; t_0, u_0) \geq \rho(t_0, u_0) > 0$$
$$u(t; t_0, u_0 + \delta_1 \bar{e}) > \rho(t_0, u_0 + \delta_1 \bar{e}) \geq \rho(t_0, u_0) + \nu \bar{e}$$

for some $\nu = \nu(\sigma_1) > 0$.

Then the assertion of Theorem 2.21 *is valid.*

The proof begins similarly to that of Theorem 2.21, but instead of (17.34) we now have

$$\dot{v}|_1 \geq F(t, v(t, x(t))) + \mu \phi(t, x(t)). \tag{17.36}$$

Now, we denote $t_s = t_0 + sT$, $x_s = x(t_s)$; $u_s(t) = u(t; t_s, v(t_s, x_s))$ is a solution of the comparison system (17.17), $s = 0, 1, \ldots,$ Let $u(t_s) \geq \alpha \bar{e}$, then it is not difficult to show that $(t, x(t)) \in \{v > 0\}$, for $t_s \leq t \leq t_{s+1}$ and sufficiently small $\mu > 0$ and that

$$u_{s+1} = v(t_{s+1}, x_{s+1}) \geq u_s(t_{s+1}) + \mu \frac{\delta}{2} \bar{e}, \tag{17.37}$$

where $T = T(\alpha) > 0$, $\delta(\alpha) > 0$ are determined from condition (e) of Theorem 2.21. To do that, we use the representation (17.26) of the solution of the auxiliary Cauchy problem

$$\dot{U} = F(t, U) + \mu \phi(t, x(t)), \quad U(t_s, \mu) = u_s. \tag{17.38}$$

Now we show that

$$v(t_s, x_s) = u_s(t_s) \geq \rho(t_0, u_0) + s\nu(\mu)\bar{e}, \tag{17.39}$$

where ρ and ν are determined by condition (b) of the theorem. Indeed, according to (17.37), $u_1(t_1) \geq u_0(t_1) + \delta_1 \bar{e}$, with $\delta_1 = \frac{1}{2}\mu\delta$. Hence, by virtue of the comparison lemma and condition (b)

$$\rho(t_1, u_1) \geq \rho(t_1, u_0(t_1) + \delta_1 \bar{e}) \geq \rho(t_1, u_0(t_1)) + \nu \bar{e} \geq$$
$$\geq \rho(t_0, u_0(t_0)) + \nu \bar{e} = \rho(t_0, u_0) + \nu \bar{e}.$$

Let (17.39) be valid for $s = k$. Let us show that this inequality holds for $s = k + 1$ as well.

$$u_k(t_k) \geq \rho(t_0, u_0) + k\nu\bar{e},$$

$$u_{k+1}(t_{k+1}) \geq u_k(t_{k+1}) + \delta_1\bar{e} \Rightarrow$$

$$\rho(t_{k+1}, u_{k+1}) \geq \rho(t_{k+1}, u(t_{k+1}) + \delta_1\bar{e}) \geq \rho(t_{k+1}, u_k(t_{k+1})) + \nu\bar{e} \geq$$

$$\geq \rho(t_k, u_k(t_k)) + \nu\bar{e} \geq \rho(t_0, u_0) + k\nu\bar{e} + \nu\bar{e} =$$

$$\rho(t_0, u_0) + (k+1)k\nu\bar{e}.$$

As $s \to \infty$, the RHS of the inequality (17.39) increases boundlessly in each component. But, by virtue of condition (d) of Theorem 2.21, for some k we have the inequality $v_k(t, x) \leq b(\| x \|)$, $b \in K$. Thus, $b(\| x \|) \geq \alpha + s\nu(\mu)$. This contradicts the assumption that $\| x(t) \|$ is bounded.

7. In the theorems on instability, the zero solution of the comparison system is assumed to be stable with respect to positive initial perturbations $u(t_0)$, as in the theorems on stability. The reason is that the instability of the comparison system (17.17) immediately leads, according to condition (c) of Theorem 2.22, to instability of the unperturbed system (17.2), and this means automatically that (17.1) is unstable.

CHAPTER 3

Stability of Systems of Ordinary Differential Equations with Quasi – Periodic Coefficients

In this chapter, the Lyapunov generalized method is used to investigate stability of systems containing a quasi – periodic temporal nonlinearity under assumption that the linearized system has imaginary characteristic exponents. Investigations are carried out by constructing a perturbed Lyapunov function, based on the Lyapunov function for linear system. The perturbation u contains a nonlinearity and is quasi – periodic in time. In calculations of the mean, the resonance terms yield forms of order two and four and a condition of asymptotic stability becomes the condition of negative definiteness of these forms.

102

3.1   Investigation of Stability by Means of Lyapunov Function of Linear System

In what follows, we consider a class of systems with nonlinearity which is quasi – periodic in time, under assumption that the matrix of linear approximation has imaginary eigenvalues. A method of reduction which reduces multi – frequency systems to the normal form is known (cf. [13] – [15]). This method is based on such change of variables in the systems that the oscillating harmonics are excluded from their RHS. The normal form of the system is simpler than the multi – frequency one, since it contains only the resonance harmonics, which makes it possible to apply Lyapunov second method in stability studies for applied problems.

The generalization of Lyapunov second method described above can be also applied to multi – frequency systems. In these cases, the stability is established by such averaging that the fast – oscillating harmonics disappear.

There exists an important class of systems where both reduction to the normal form ([68]) and the generalized Lyapunov method are applicable: these are systems with quasi – periodic temporal nonlinearity. Here we present some new results obtained with the help of generalized Lyapunov method.

Let us consider a system with almost periodic nonlinearity ([42]):

$$\dot{x}_j = \imath\lambda_j x_j + \mu F_j(x, t),$$

$$(1.1)$$

$$\dot{x}_{j+n} = -\imath\lambda_j x_{j+n} + \mu F_{j+n}(x, t),$$

where $\mu > 0$ is a small parameter, $\imath^2 = -1$, $j = 1, \ldots, n$, $\imath\lambda_j$ and $-\imath\lambda_j$ are the eigenvalues of the linear approximation matrix. The functions $F_j(x, t)$ are expanded into the Taylor – Fourier series

$$F_j(x, t) = F_{j,k}(x, t) + F_{j,k+1}(x, t) + \ldots, \quad k \geq 1, \quad j = \overline{1, 2n},$$

where

$$F_{j,\bullet}(x.t) = \sum_{|Q|=\bullet} F_{j,\alpha}(t)X^Q, \quad X^Q = x_1^{q_1}, \ldots, x_{2n}^{q_{2n}},$$

$$Q = (q_1, \ldots, q_{2n}), \quad |Q| = q_1 + \cdots + q_{2n}, \quad 0 \leq Q \in \mathbf{Z}^{2n},$$

$\mathbf{Z}^{2n}$ is an integer lattice in $\mathbf{R}^{2n}$, $F_{jQ}(t) = \sum f_{jQP} \exp \imath < P, \Omega > t$, $f_{jQP} \in \mathbf{C}$, $P \in \mathbf{Z}^m$ $\Omega = (\omega_1, \ldots, \omega_m) \in \mathbf{R}^m$, $< P, \Omega > = \sum_{i=1}^m P_i \omega_i$.

The series $\sum f_{jQP}$ converges absolutely. The system (1.1) is assumed to be obtained by a non − degenerate linear transformation from a certain real system, hence the quantities x_j and x_{j+n} are complex conjugated. Systems of the form (1.1) result from reduction of nonlinear system with a constant linear approximation matrix which has only imaginary eigenvalues, all of which are different or multiple with simple elementary divisors, as well as with a non − constant matrix, reducible to a constant one with indicated properties.

The necessary conditions for stability in linear systems of the type (1.1) related to parametric resonance are known (cf. [32]).

Let us introduce the notation

$$\Lambda = (\lambda_1, \ldots, \lambda_n, -\lambda_1, \ldots, -\lambda_n),$$

$$\Theta = (\theta_1, \ldots, \theta_n, -\theta_1, \ldots, -\theta_n) \in \mathbf{R}^{2n}, \quad \theta = (\theta_1, \ldots, \theta_n), \quad (1.2)$$

$$r = (r_1, \ldots, r_n) \in \mathbf{R}^n, \quad R^Q = r_1^{q_1+q_{1+n}} \cdots r_n^{q_n+q_{2n}}.$$

The system (1.1) is said to have a resonance, if for some $j = 1, \ldots, n$ $0 \leq Q \in \mathbf{Z}^{2n}$, $P \in \mathbf{Z}^m$, we have

$$< P, \Omega > + < Q, \Lambda > -\lambda_j = 0. \tag{1.3}$$

If the relation (1.3) holds also for Q such that $< Q, \Lambda > \neq \lambda_j$, the eigenvalues are said tio be linked with the frequency basis Ω of the function $F_{jQ}(t)$.

The system (1.1) has the *inner resonance* of order l, if for some $j = 1, \ldots, n$, $0 \leq Q \in \mathbf{Z}^{2n}$, $|Q| = l$, we have

$$< Q, \Lambda > -\lambda_j = 0. \tag{1.4}$$

An inner resonance is called identical, if Eq. (1.4) is true for all $\Lambda \in \mathbf{R}^{2n}$ for some Q_P.

An identical resonance has odd order only.

The vectors P and Q satisfying Eqs. (1.3) and (1.4) are called resonance vectors.

A resonance vector, different from Q_P will be denoted as $\tilde{Q}$. Identical resonances have, as in autonomous systems [90], a decisive effect on the nature of stability.

Consider the truncated system

$$\dot{x}_j = \imath\lambda_j x_j,$$

$$\dot{x}_{j+n} = -\imath\lambda_j x_{j+n}, \tag{1.5}$$

where $j = \overline{1,n}$. This system has the first integral

$$v_0(x) = x_1 x_{1+n} + \cdots + x_n x_{2n}.$$

The solution of the system (1.5) can be conveniently expressed in the trigonometrical form, to wit

$$\bar{x}_j(t) = r_j \exp[\imath(\lambda_j t + \theta_j)],$$

$$\bar{x}_{j+n}(t) = r_j \exp[-\imath(\lambda_j t + \theta_j)], \quad j = \overline{1,n}. \tag{1.6}$$

Let us introduce the notation

$$\phi_0^R(x,t) = \sum_j x_{j+n} F_{jk}(x,t), \quad \phi_0(x,t) = 2\mathrm{Re}\phi_0^R(x,t). \tag{1.7}$$

Then, with accuracy up to the terms of order greater than $k + 1$ with respect to x, the total derivative of the function $v_0(x)$ is, by virtue of the system (1.1), equal to

$$\dot{v}_0 = \mu\phi_0(x,t) + \ldots.$$

We introduce the mean $\psi(r, \theta)$ along the solutions (1.6) of the truncated system.

$$\psi(r, \theta) = \underset{t}{M}\{\phi_0(\bar{x}(t, r, \theta), t)\} =$$

$$= \lim_{T \to \infty} T^{-1} \int_0^T \phi_0(\bar{x}(t, r, \theta), t)\, dt. \tag{1.8}$$

Since here $\phi_0(\bar{x}, t)$ is an almost periodic function, the mean $\psi(r, \theta)$ always exists.

Let $\bar{x}(t, x_0)$ be the solution of the linear system (1.5) with the initial condition $\bar{x}(0) = x_0$.

In the polar coordinates r, θ we have

$$\psi(x_0) = \psi(r, \theta). \tag{1.9}$$

Let us also introduce the notation

$$R = (r_1, \ldots, \dot{r}_n, \dot{r}_1, \ldots, \dot{r}_n), \quad Q_j = Q + E_{j+n},$$

with the unit vector E_{j+n}. Using this notation, we obtain the expression for the mean

$$\psi^R(r, \theta) = \sum_{j=1}^n \sum_{|Q|=k} \underset{t}{M}\{F_{jQ}(t)e^{\imath<Q_j,\Lambda>t}\}e^{\imath<Q_j,\theta>t}R^Q. \tag{1.10}$$

We note that $a_{jQ} = \underset{t}{M}\{F_{jQ}(t)e^{\imath<Q_j,\Lambda>t}\}$ is a value of the spectral function at the point $- < Q_j, \Lambda >$ of almost periodic function F_{jQ}. Thus, the non – zero contribution to the mean $\psi^R(r, \theta)$ comes from those terms, for whose the number $- < Q_j, \Lambda >$ belongs to the spectrum of the function $F_{jQ}(t)$. Accordingly, we introduce the class ξ_{jk} of the vectors Q, relevant for this system

$$\xi_k = \{Q : 0 \leq Q \in \mathbf{Z}^{2n}, \ |Q| = k, \ a_{jQ} \neq 0\}. \tag{1.11}$$

Let us introduce the notation

$$A_{jQ} = 2\mathrm{Re}[a_{jQ} \exp(\imath < Q_j, \Theta >)]. \tag{1.12}$$

Thus, the mean ψ is a homogeneous polynomial (a form) of order $k+1$ in $r_1, \ldots, r_k$:

$$\psi(r, \theta) = \sum_{j=1}^{n} \sum_{Q \in \mathcal{E}_{k,j}} A_{jQ}(\theta) R^{Qj}, \qquad (1.13)$$

and stability is established by examining the sign of this form.

Theorem 3.1 *If the form* (1.13) *is negative – definite (or positive – definite) with respect to* $r_1, \ldots, r_n$, *then for* $k > 1$ *the equilibrium position of the system* (1.1) *is asymptotically stable (unstable).*

Remark. Smallness of the parameter μ is important only in the case when $k = 1$. For $k > 1$, the parameter μ may take any positive value.

Proof. The asymptotic stability is established by applying Theorem 2.1. It is sufficient to check uniformity of the limit transition (1.8) with respect to the initial conditions of the linear system.

The function $\psi(x_0) = \psi(r, \theta)$ is a negative – definite form of order $K + 1$ with respect to $r_1, \ldots, r_n$, therefore

$$\psi(r, \theta) \leq -\eta \parallel r \parallel^{k+1},$$

$$\eta = \min_{\parallel r \parallel = 1} |\psi(r, \theta)|, \quad \parallel r \parallel = r_1 + \cdots + r_n.$$

Next, we examine the function

$$k(\Delta t, x_0) = \frac{1}{\Delta t} \int_0^{\Delta t} \phi(\bar{x}(t, x_0), t)\, dt - \psi(x_0).$$

Thus function is also a form of order $K + 1$ with respect to $r_1, \ldots, r_n$ and it can be represented as a sum of monomials of the form

$$2\mathrm{Re}[(\frac{1}{\Delta t} \int_0^{\Delta t} F_{jQ}(t) \exp \imath < Q_j, \Lambda > t\, dt - a_{jQ}) \exp \imath < Q_j, \theta > R^Q.$$

There exists a majorant $Q_0(\Delta t)$ such that $|k| < Q_0(\Delta t) \parallel r \parallel^{k+1}$, where $Q_0 \to 0$ as $\Delta t \to 0$ and there exists a constant l_0 such that $|k(\Delta t, x_0)| \leq \frac{1}{4}|\psi(x_0)|$ for $\Delta t > l_0$ and $\parallel x \parallel \leq h$.

The corresponding assertion regarding instability is proved with the help of the generalized Lyapunov second method given in Ch. 2, and by making use of estimates presented in Ref. [5].

In order to transform an expression for the mean (1.10), we introduce the notation

$$\bar{\psi}(r) = \sum_{j=1}^{n} \sum_{Q_p \in \xi_{kj}} A_{jQ_P} R^{Q_{Pj}}, \tag{1.14}$$

$$\tilde{\psi}(r, \theta) = \sum_{j=1}^{n} \sum_{\tilde{Q} \in \xi_{kj}} A_{j\tilde{Q}}(\theta) R^{\tilde{Q}_j} \tag{1.15}$$

Here, $\tilde{Q}$ is a resonance vector which is not identically resonant. As it is easy to see from (1.12), at $Q = Q_P$, A_{jQ} is independent of the vector of initial phase θ:

$$A_{jQ} = 2\mathrm{Re} \underset{t}{M} \{F_{jQ_p}(t)\}.$$

According to (1.13) – (1.15), we have

$$\psi(r, \theta) = \bar{\psi}(r) + \tilde{\psi}(r, \theta). \tag{1.16}$$

It follows that for even k, when the identical resonances of order K are absent, the mean $\bar{\psi}(r) \equiv 0$ changes the sign, which, as a rule, leads to instabilities.

When K is odd, the mean $\psi(r, \theta)$ may have a constant sign if $\bar{\psi}(r)$ has a constant sign and suppresses $\tilde{\psi}(r, \theta)$, whose sign varies.

Below, we consider a case where for any Q, $|Q| = k$, and j, $1 \leq j \leq n$, the spectrum of the function $F_{jQ}(t)$ does not contain numbers $< Q_j, \Lambda > \neq 0$, and the system has identical inner resonances. Moreover,

$$\psi(r, \theta) = \bar{\psi}(r). \tag{1.17}$$

It can be easily checked that for $k = 2s + 1$, $s = 0, 1, 2, \ldots$, $\bar{\psi}$ is a form of order $s + 1$, with respect to $r_1^2, \ldots, r_n^2$.

Let $k = 1$. This means that in the system (1.1) the leading term in perturbations $F_j(x,t)$ is linear in x. From the point of view of the considered method, for investigating stability, the systems in standard form are especially convenient, since in this case the truncated system is trivial and there is the widest choice of Lyapunov and Chetayev functions.

The system (1.1) is reduced to the standard form by the substitution

$$x = \Phi(t)u, \qquad (1.18)$$

where

$$\Phi(t) = \text{diag}[e^{i\lambda_1 t}, \ldots, e^{i\lambda_n t}, e^{-i\lambda_1 t}, \ldots, e^{i\lambda_n t}].$$

In terms of u, we obtain

$$\dot{u}_j = \mu G_j(u,t) + o(\|u\|^k), \quad j = 1, \ldots, 2n, \qquad (1.19)$$

where

$$G_j(u,t) = \sum_{|Q|=k} G_{jQ}(t)U^Q, \quad G_{j+n}(u,t) = \overline{G_j(u,t)},$$

$$G_{jQ}(t) = F_{jQ}(t)e^{i<Q_j,\Lambda>t}, \quad U = (u_1, \ldots, u_n).$$

For $k = 1$, the system (1.19) is linear in the first approximation

$$\dot{u}_j = \mu[G_{jE_1}u_1 + \cdots + G_{jE_{2n}}u_{2n}] + o(\|u\|^k), \quad j = 1, \ldots, 2n; \quad (1.20)$$

$$G_{jE_s}(t) = F_{jE_s}(t)\exp(i(\lambda_c - \lambda_j)t). \qquad (1.21)$$

We average the system (1.20) over the time variable which enters the system explicitly:

$$\dot{\xi}_j = \mu \sum_{s=1}^{2n} \underset{t}{M}\{G_{jE_s}(t)\}\xi_s, \quad j = 1, \ldots, 2n. \qquad (1.22)$$

For non – critical cases the nature of stability is determined by the properties of the linear system (1.22) with constant coefficients. Let us denote by P the matrix of coefficients of the linear system. We will prove the following

Theorem 3.2 *If all the roots of the characteristic equation*

$$\det(P - \lambda E) = 0$$

of the averaged system (1.22) have negative real part (or at least one root has positive real part), then there exists $\mu_0 > 0$ such that for all μ, $0 < \mu \leq \mu_0$ and $k = 1$, the equilibrium position $x = 0$ of the system (1.1) is asymptotically stable (respectively, unstable).

The proof of this theorem uses the technique of constructing a Lyapunov function for linear system with constant coefficients ([75]) and the generalized Lyapunov method presented in Chapter 2.

The result of Theorem 3.2 is one of the simplest and is closely related to the already known ones (cf. [11]).

We examine the case $k = 3$, i.e. we assume that the system is nonlinear. We also assume the the relation (1.17) holds, and moreover that the mean $\psi(r,t) = \psi(r)$ is a quadratic form in non – negative variables $r_1^2, r_2^2 \ldots, r_n^2$,

$$\psi(r) = \sum_{j=1}^{n} r_j^2 \sum_{s=1}^{n} A_{j,s} r_s^2, \quad A_{j,s} = A_{j,E_j + E_s + E_{s+m}}. \tag{1.23}$$

The question, as to whether the fixed sign of this form leads to stability is answered by Theorem 3.1. The necessary and sufficient conditions for the quadratic form to have a fixed sign in the non – negative orthant are given in Refs. [33], [64]. The solution for a symmetric matrix $A_{j,s}$ is given by the following theorem.

Theorem 3.3 *Let $A_{j,s} = A_{s,j}$, for $j, s = 1, \ldots, n$ and let the form (1.23) take positive values. Then the system (1.1) is unstable with respect to x for sufficiently small μ.*

The proof is given in Ref. [62].

3.2   Construction of Perturbed Lyapunov Function for Higher Order Resonances

In order to study resonances of order four, we consider a more general system of differential equations, which is almost periodic in time and contains polynomials of second, third, and higher orders in the unknown variable z ([58])

$$\dot{z} = \Lambda z + f_2(t, z) + f_3(t, z) + f_4(t, z), \qquad (2.1)$$

where by z we denote the set of variables $x_1, y_1, \ldots x_n, y_n$.

For a system of linear approximation in k variables ($k \leq n$), we choose the Lyapunov function in the following form

$$v_0(z, t) = \sum_{j=1}^{k} |\lambda_j| \frac{x_j^2 + y_j^2}{2}. \qquad (2.2)$$

We construct the perturbed Lyapunov function as a truncated series in powers of $z = (x,, y)$, $v = v_0 + s$, where the perturbation s is a homogeneous polynomial in z, constructed in such a way that the total derivative of v, by virtue of the system (2.1), starts with terms of even order in z

$$\dot{v} = \frac{\partial v_0}{\partial z} \Lambda z + \frac{\partial v_0}{\partial z} f_2 + \frac{\partial s}{\partial t} + \frac{\partial s}{\partial z} \Lambda z + \frac{\partial v_0}{\partial z} f_3 + \frac{\partial v_0}{\partial z} f_4 +$$

$$+ \frac{\partial s}{\partial z} f_2 + \frac{\partial s}{\partial z} f_3 + \frac{\partial s}{\partial z} f_4. \qquad (2.3)$$

Let us introduce the notations

$$\psi(t, z) = \frac{\partial s}{\partial z} f_2 + \frac{\partial v_0}{\partial z} f_3,$$

$$H(t, z) = \frac{\partial v_0}{\partial z} f_2, \qquad (2.4)$$

$$\phi(t, z) = \frac{\partial s}{\partial z}(f_3 + f_4) + \frac{\partial v_0}{\partial z} f_4.$$

Let the perturbation of Lyapunov function satisfy the equation

$$\frac{\partial s}{\partial t} + \frac{\partial s}{\partial z}\Lambda z + H = 0. \tag{2.5}$$

We expand the function H into a series in complex variables $u_j = x_j + \imath y_j$, $v_j = x_j - \imath y_j = \bar{u}_j$ and denote by $\alpha_{m_1,\dots,k_n}(t)$ the coefficient of the term $u_1^{m_1}v_1^{k_1}\cdots u_n^{m_n}v_n^{k_n}$ (m_i and k_i are natural numbers):

$$H(x,y,t) = \sum_{m_1+\cdots+k_n=3} \alpha_{m_1,\dots,k_n}(t)u_1^{m_1}\cdots v_n^{k_n}. \tag{2.6}$$

Accordingly, $s(x,y,t)$ is represented in the form:

$$s(x,y,t) = \sum_{m_1+\cdots+k_n=3} \beta_{m_1,\dots,k_n}(t)u_1^{m_1}\cdots v_n^{k_n}. \tag{2.7}$$

The characteristics of the system (2.5) are integral curves of the system (1.5), written in the form (1.6).

We set

$$u_j = r_j \exp \imath(\lambda_j t - \theta), \quad v_j = r_j \exp -\imath(\lambda_j t + \theta). \tag{2.8}$$

Since the coefficients α and β are interrelated by linear differential equation

$$\dot{\beta}_{m_1,\dots,k_n} + \imath[\lambda_1(k_1 - m_1) + \cdots + \lambda_n(k_n - m_n)]\beta_{m_1,\dots,k_n} =$$
$$= \alpha_{m_1,\dots,k_n}(t) \tag{2.9}$$

and the imaginary eigenvalues are not subject to resonance relations up to third order inclusive, we have

$$\beta_{m_1,\dots,k_n} = \exp\{-\imath(\lambda_1(k_1 - m_1) + \cdots + \lambda_n(k_n - m_n))t\} \times \tag{2.10}$$

$$[C_{m_1,\dots,k_n} - \int_0^t \alpha_{m_1,\dots,k_n}(t)\exp\{-\imath(\lambda_1(k_1 - m_1) + \cdots + \lambda_n(k_n - m_n))t\}\,dt].$$

Let us calculate $\bar{\psi}(r,\theta)$, the mean of order four along the solution (2.8) taken over the almost periodic function $\psi(r,\theta,t)$,

$$\psi(r,\theta,t) = \sum_{Q}\sum_{m} \psi_{Qm} r^Q \exp\{\imath(<Q,\Lambda>t + \Omega t + <Q,\Theta>)\}. \quad (2.11)$$

The mean $\bar{\psi}$ contains resonance terms only

$$\bar{\psi}(r,\theta) = \sum_{Q}\sum_{m} \psi_{Qm} r^Q \exp\{\imath <Q,\Theta>\}, \quad <Q,\Lambda> +\Omega = 0.$$

As in the normal form techniques ([18], [13]), stability is determined by resonance harmonics only.

Theorem 3.4 *Let the RHS of the system* (2.1) *consist of functions almost periodic in time and polynomial in* x, y *with imaginary eigenvalues and not subject to the resonance relations up to the third order inclusive. Let the mean* $\bar{\psi}(r,\theta)$ *be negative – definite (or positive – definite) function in part of variables* x_1, y_1, $\ldots$, x_k, y_k. *Then the equilibrium position of the system* (1.1) *is asymptotically stable (unstable).*

The proof is similar to that of Theorem 2.15.

Regarding instability, a theorem of Chetayev type ([58]) holds.

Resonances of higher orders can be dealt with using further approximations of the generalized Lyapunov function.

CHAPTER 4

Stability of Multi – Frequency Systems

In this chapter we consider multi – frequency systems. Theorems of the generalized Lyapunov second method are applied to investigate stability of the equilibrium positions of averaged systems. In course of these investigations we make use of asymptotic stability of an averaged system. We introduce the condition of isolation of resonances, which means that a resonance frequency differs from zero outside some neighbourhood of the equilibrium position. Thus, resonance harmonics oscillate outside this neighbourhood. This gives an effective estimate of small denominators, which makes use of the essential nonlinearity of the system (in this context the nonlinearity means that the system depends on 'slow' variables). Examples of systems with various numbers of slow variables and frequencies are considered. Stability is examined on asymptotically large and infinite intervals.

As examples of applications of the proven theorems, we consider a problem of motion of a satellite with respect to the centre of mass and a problem of stabilization of the upper position of equilibrium of a physical pendulum by varying the suspension point according to the harmonic low, with resonance of the eigenfrequency of the pendulum and the frequency of oscillations of the suspension point. The problem of satellite motion relative to the centre of mass is also considered when the resonance between the satellite rotation along the orbit and the frequency of its oscillations with respect to the centre of mass occurs.

4.1 Statement of the Problem

In Chapter 1 we considered multi – frequency systems of ordinary differential equations in order to study evolution by means of averaging. In what follows we study such systems with additional assumption that the averaged system has a position of equilibrium. In investigation of stability of systems of this class, the generalized Lyapunov method is very effective, since it is precisely the method which makes it possible to detect implicit dissipation in the system which exhibits itself by the fact that the mean of derivative of the perturbed Lyapunov function becomes negative – definite. The perturbation of Lyapunov function in this case is constructed so as to eliminate fast oscillating harmonics in the expression for the derivative of the perturbed Lyapunov function calculated by virtue of the initial system.

A greater difficulty in multi – frequency systems is caused by resonance harmonics which result in appearance of small denominators in the process of applying any version of perturbation theory. Below, we suggest a technique to estimate resonance frequencies ([62], [44] [9]). This technique is based on accounting for certain properties of frequencies in the neighbourhood of a point tested for stability: this requires a uniform increase of the modulus of the resonance frequency as it moves away from the resonance point.

Let us consider a multi – frequency system

$$\dot{x} = \mu X(x, q), \quad \dot{q} = \omega(x) + \mu \Phi(x, q), \qquad (1.1)$$

where $\dim x = \dim X = s$, $\dim q = \dim \omega = \dim \Phi = m$, the functions X

and Φ are periodic in q (with the period 2π) and are expandable into Fourier series which converges absolutely and uniformly in $q \in [0, 2\pi]$, $\parallel x - x_0 \parallel \leq H$. The RHS of this system satisfy the conditions of existence and uniqueness of solutions.

Let us average the RHS of the x equation in the system (1.1) over the phases q (we take zero harmonics of the Fourier expansion) and examine the averaged system

$$\dot{x} = \mu X_0(x). \tag{1.2}$$

The restrictions imposed on the convergence rate of the Fourier series for the function $X(x, q)$, and consequently on smoothness of this function, is determined by the properties of the averaged system (1.2).

The unperturbed system has in our case the form

$$\dot{x} = 0, \quad \dot{q} = \omega = \text{const.} \tag{1.3}$$

Thus, in investigating the stability of the system (1.1)by means of the theorems of Chapter 2, the function J is calculated by integrating along the integral curves of the system (1.3).

The point x_0 is said to be *the resonance point* of the system (1.1), if for any integer vector $k = \{k_1, \ldots, k_m\}$ we have the relation

$$k\omega(x_0) = 0. \tag{1.4}$$

Here, we assume that all components of the vector $\omega(x)$ differ from zero at the point x_0, i.e. $\omega_r(x_0) \neq 0$, $1 \leq r \leq m$.

Depending on the properties of the vector $\omega(x)$, the relation (1.4) may define an isolated resonance point, a resonance surface or a resonance curve. We will investigate stability of the resonance point (or surface), assuming that x_0 is asymptotically stable or is a stable equilibrium position of the averaged system (1.2), i.e.

$$X_0(x_0) = 0. \tag{1.5}$$

The systems (1.1) and (1.2) will be considered for $|x - x_0| \leq H$.

The point x_0 (or some resonance manifold) will be called (x, μ) – *stable*, if for any $\epsilon > 0$ $(\epsilon < \epsilon_0)$ there exist $\eta(\epsilon)$ and $\mu_0(\epsilon)$ such that the x component of the solution with the initial data lying in the η-neighbourhood of the point x_0 (resonance manifold) will not leave the ϵ-neighbourhood of x_0 for $\mu < \mu_0(\epsilon)$ for all subsequent t.

4.2 Stability of Single – Frequency Systems of Equations with Asymptotically Stable Averaged System

Let us first examine a single – frequency case (1.1); the case of fast rotating phase ([11], [87]) which reduces to the standard – form system

$$\dot{x} = \mu X(x, t). \tag{2.1}$$

This single – frequency system does not contain resonances and the RHS of this system admits averaging over the explicit time

$$X_0(x) = \lim_{T \to \infty} \frac{1}{T} \int_{t_0}^{t_0+T} X(x, t) \, dt. \tag{2.2}$$

Theorems on averaging on an infinite interval for system (2.1) was proved in [11] under assumption that the averaged system

$$\dot{x} = \mu X_0(x) \tag{2.3}$$

has a quasi – static or periodic asymptotically stable solution with stability ensured by negativeness of the real parts of the characteristic exponents of equations in variations of averaged system.

Let us investigate stability of a quasi – static solution x_0 of the averaged system, assuming that the asymptotic stability of this solution is ensured by existence of a positive – definite Lyapunov function with negative – definite derivative. The corresponding theorem is formulated for systems in the standard form (2.1).

Theorem 4.1 *Suppose that:*

a) The mean (2.1) exists uniformly in $t_0 > 0$, $\| x - x_0 \| \leq H$.

b) The averaged system (2.3) has a positive − definite Lyapunov function $v_0(x)$ which has a negative − definite derivative $\mu \nabla v_0 X_0(x)$ for $t > 0$, $\| x - x_0 \| \leq H$.

c) There exist summable functions $M(t)$, $F(t)$ and constants F_0, M_0 as well as a non − decreasing function $\chi(\alpha)$, $\lim_{\alpha \to 0} \chi(\alpha) = 0$ such that for $t > 0$, $\| x - x_0 \| \leq H$ we have

$$\| \nabla v_0(x') X_0(x', t) - \nabla v_0(x'') X_0(x'', t) \| \leq \chi(|x' - x''|) F(t),$$

$$\int_{t_1}^{t_2} F(t)\, dt \leq F_0(t_2 - t_1),$$

$$\| X(x, t) \| < M(t), \quad \int_{t_1}^{t_2} M(t)\, dt \leq M_0(t_2 - t_1)$$

on any finite interval $[t_1, t_2]$.

Then for any $\epsilon > 0$ there exist $\eta(\epsilon)$ and $\mu_0(\epsilon)$ such that any solution $x = x(t)$ of the system (2.1) with initial condition $x(0) = \bar{x}_0$ with $\| x_0 - \bar{x}_0 \| < \epsilon$ for $\mu < \mu_0(\epsilon)$ and for all $t > 0$ satisfies the inequality $\| x(t) - x_0 \| < \epsilon$.

Proof. Let us examine the Lyapunov function $v_0(x)$ of the averaged system. This function satisfies condition (a) of Theorem 2.5. Since $f(t, z) \equiv 0$ for the system (2.1), and since the function $X(x, t)$ plays a role of the perturbation, condition (b) of Theorem 2.5 is satisfied as well ($\nabla v_0 f \equiv 0$). This means that we are dealing with the 'neutral' case.

Let us single out oscillating terms on the RHS of (2.1) and denote them $\tilde{X}(x, t)$

$$X(x, t) = X_0(x) + \tilde{X}(x, t). \tag{2.4}$$

This separation is possible because condition (b) of the theorem has been proved.

Now we differentiate $v_0(x)$ by virtue of the system (2.1), taking (2.4) into account:

$$\frac{dv_0}{dt} = \nabla v_0 X_0(x) + \nabla v_0 \tilde{X}_0(x, t). \tag{2.5}$$

Since $\nabla v_0 X_0(x)$ is negative – definite and by virtue of condition (b) of the theorem, conditions (c) are satisfied.

The smoothness conditions, i.e. conditions (d) of Theorem 2.5 are also fulfilled since the conditions of Theorem 4.1 are identical. Thus, we have satisfied all the conditions of Theorem 2.5, and hence Theorem 4.1 is valid.

4.3 Stability of Multi – Frequency Systems of Equations with Asymptotically Stable Averaged System

1. Now let us consider the system (1.1) in the multi – frequency case (with m frequencies). As above, we assume that the averaged system (1.2) has the asymptotically stable point x_0, in the neighbourhood of whose there exists a Lyapunov function $v_0(x)$ which is positive – definite and has a negative – definite derivative.

We fix $\epsilon > 0$ and set

$$v_\epsilon = \min v_0(x) \quad \text{for} \quad \| x - x_0 \| = \epsilon. \tag{3.1}$$

Then the surface $v_0(x) = v_\epsilon$ belongs to the ball $\| x - x_0 \| \leq \epsilon$. We also set

$$\gamma_0(\epsilon) = \min \| x - x_0 \| \quad \text{for} \quad v_0(x) = v_\epsilon. \tag{3.2}$$

By virtue of the properties of the Lyapunov function, for any γ, $0 < \gamma < \gamma_0$, the inequality

$$v_0(x) < v_\epsilon \quad \text{for} \quad \| x - x_0 \| < \gamma \tag{3.3}$$

holds.

Resonances may occur in multi – frequency systems, i.e. the combinative frequencies are nullified

$$\lambda_k(x) = k\omega(x), \tag{3.4}$$

where k is an integer – valued vector.

Let us denote by $K(r_0)$ the set of integer – valued vectors such that

$$\lambda_{k_r}(x_0) = 0, \tag{3.5}$$

where $|k_r| < r_0$, r_0 being a finite number.

Definition 4.1 *A resonance at the point x_0 is called r_0 – isolated if for any ρ_1 and any k_r satisfying the condition (3.5) there exists $\Omega_1 > 0$ such that, if*

$$\| x - x_0 \| > \rho_1 \quad then \quad |\lambda_{k_r}(x)| > \Omega_1. \tag{3.6}$$

The sum of the corresponding terms of the series $X(x, q)$ will be denoted by

$$X_{r_0}(x, q) = \sum_{|k_r < r_0} X_{k_r}(x)e^{i(kq)}.$$

We consider now a monotonically increasing function $\Omega(\epsilon)$, $\Omega(0) = 0$ and denote by $\tilde{K}_\epsilon(\Omega)$ the set of vectors $\tilde{k}$ which satisfy the inequality

$$\| \tilde{k}\omega(x) \| \geq \Omega(\epsilon) \tag{3.7}$$

for all x such that $0 < |x - x_0| < \epsilon < \epsilon_0$. The sum of the corresponding terms of the series $X(x, q)$ will be denoted by $\tilde{X}_\epsilon(x, q, \Omega)$.

We introduce the additional notation:

$\tilde{K}(r_0, \Omega)$ is a union of the sets $\tilde{K}_\epsilon(\Omega)$ and $K(r_0)$; $\bar{K}_\epsilon(r_0, \Omega) = K - \tilde{K}(r_0, \Omega)$ is the complement of $\tilde{K}(r_0, \Omega)$ in K, which is the set of all integer – valued vectors with non – zero coefficients in the Fourier expansion of the function $X(x, q)$;

$$\tilde{X}(x, q) = X_{r_0}(x, q) + \tilde{X}_\epsilon(x, q, \Omega),$$

$\bar{X}_\epsilon(x, q)$ is a sum of terms from the series for $X(x, q)$ corresponding to $\bar{K}_\epsilon(r_0, \Omega)$; $R_\epsilon(x, q, \Omega)$ is a sum of moduli of the terms of the series for $X(x, q)$ corresponding to $\bar{K}_\epsilon(r_0, \Omega)$.

In this chapter, we suggest a method of estimating small frequencies (denominators), based on separation of them according to their magnitude with regard to the sizes of the ϵ-neighbourhood of the point being tested for stability, i.e. taking into account the location of trajectories in the variables x. We will single out the following three classes of frequencies (denominators) and the corresponding harmonics of the series of $X(x, q)$:

1. Resonances whose stability is tested and whose frequencies (denominators) differ from zero outside certain neighbourhood of the resonance.

2. Frequencies which are estimated by the quantity $\Omega(\epsilon)$ and the corresponding oscillating terms.

3. Resonance harmonics whose frequencies cannot be estimated from below (these harmonics, grouped in $\bar{X}_\epsilon(x, q)$ will have to be estimated by the modulus).

We assume that the following conditions are satisfied.

A. *The averaged system* (1.2) *has a positive – definite Lyapunov function* $v_0(x)$ *whose derivative*

$$\dot{v}_0 = \mu \nabla v_0 X_0(x)$$

is negative definite for $\| x - x_0 \| \leq H$.

B. *The functions* $X(x, q)$ *are expanded into Fourier series in* q *which converge absolutely and uniformly in* $q \in [0, 2\pi]^m$, $\| x - x_0 \| \leq H$. *Let the function* $\nabla v_0 X(x, q)$ *satisfy the following smoothness conditions in the variable* x: *there exists a non – decreasing function* $\chi(\alpha)$, $\lim_{\alpha \to 0} \chi(\alpha) = 0$ *such that for* $\| x - x_0 \| \leq H$, $q \in [0, 2\pi]^m$ *we have*

$$\| \nabla v_0(x') X_0(x', q) - \nabla v_0(x'') X_0(x'', q) \| \leq \chi(|x' - x''|).$$

Theorem 4.2 *Suppose that conditions A and B are fulfilled. Let the resonance at the point x_0 be r_0 – isolated for some r_0. Moreover, let there exist functions $\Omega(\epsilon)$ and $\gamma(\epsilon)$, $(0 < \gamma(\epsilon) < \gamma_0(\epsilon))$ such that for $\gamma(\epsilon) \leq \| x - x_0 \| \leq \epsilon$ the inequality*

$$\max_{\gamma(\epsilon) \leq \|x-x_0\| \leq \epsilon} \| \nabla v_0 \| R_\epsilon(x, r_0, \Omega) < -\frac{1}{2} X_0(x) \nabla v_0(x). \qquad (3.8)$$

holds.

Then for any $\epsilon > 0$ $(0 < \epsilon < \epsilon_0 < H)$ there exist $\eta(\epsilon)$ and $\mu_0(\epsilon)$ such that any solution $x = x(t)$, $q = q(t)$ of the system (1.1) with the initial data $x(0) = \bar{x}_0$, $q(0) = q_0$, $\| x - x_0 \| < \eta(\epsilon)$, for $\mu < \mu_0(\epsilon)$, satisfies the inequality $\| x(t) - x_0 \| \leq \epsilon$ for all $t > 0$.

Proof. In order to investigate stability of the resonance point x_0, we apply Theorem 2.1, setting $u \equiv 0$. Then $v = v_0$ and the derivative of the function $v_0(x)$ has, by virtue of (1.1), the form

$$\frac{dv}{dt} = \mu \frac{\partial v_0}{\partial x} X(x, q) = \mu \phi(x, q). \qquad (3.9)$$

Condition (a) of Theorem 2.1 is ensured by existence of the Lyapunov function v_0 of the averaged system; condition (b) is satisfied since x's are slow variables in the initial system, i.e. $f \equiv 0$; the smoothness condition (d) is fulfilled by virtue of condition B. of Theorem 4.2; finally, the condition (e) is fulfilled by the choice $u \equiv 0$.

The only condition, which remains to be checked is therefore condition (c) of Theorem 2.1. Having fixed $\epsilon > 0$, let us divide the vectors k into the sets $\tilde{K}_\epsilon(r_0, \Omega)$, $K(r_0)$, $\bar{K}_\epsilon(r_0, \Omega)$, as indicated above and represent $\phi(x, q)$ in the form

$$\phi(x, q) = \nabla v_0 X_0(x) + \nabla v_0 \bar{X}_\epsilon(x, q) + \nabla v_0 X_{r_0}(x, q) +$$
$$+ \nabla v_0 \tilde{X}_\epsilon(x, q, \Omega). \qquad (3.10)$$

We consider the non – oscillating terms

$$\nabla v_0 X_0(x) + \nabla v_0 \bar{X}_\epsilon(x, q) \qquad (3.11)$$

and introduce the notation

$$\min_{0 < k \le |x - x_0| \le \epsilon} [-\nabla v_0 X_0(x)] = B_k. \qquad (3.12)$$

By virtue of condition A, for any η which satisfies the inequality

$$\gamma < \eta < \gamma_0(\epsilon), \qquad (3.13)$$

we get

$$B_\gamma \le B_\eta. \qquad (3.14)$$

By virtue of condition (3.8), we have

$$\max \| \nabla v_0 \| \, \tilde{R}_\epsilon(x, r_0, \Omega) < \frac{1}{2} B_\gamma. \qquad (3.15)$$

Hence, introducing the notation $B_\eta - \frac{1}{2} B_\gamma = \delta$, with $\delta > 0$, for $\eta <$ $\| x - x_0 \| < \epsilon$, we obtain the estimate

$$\nabla v_0 X_0(x) + \nabla v_0 \bar{X}_\epsilon(x, q) < -\delta. \qquad (3.16)$$

According to (3.7), the frequencies of oscillations entering $\tilde{X}_\epsilon(x, q, \Omega)$ are bounded by $\Omega(\epsilon)$ from below. By virtue of assumptions of the theorem with regards to the function $\Omega(\epsilon)$, for $\eta < \| x - x_0 \| < \epsilon$ one can choose $\Omega_1 = \Omega_1(\epsilon)$ such that all the frequencies entering $X_{r_0}(x, q)$ are estimated from below by $\Omega_1(\epsilon)$. Therefore, there exists $l(\epsilon)$ such that

$$\int_{t_0}^{t_0+T} [\tilde{X}_\epsilon(x, q, \Omega) + X_{r_0}(\bar{x}, \bar{q})] \cdot \nabla v_0 \, dt < \frac{1}{4} \delta T \qquad (3.17)$$

for $T > l(\epsilon)$ and any t_0, where the integration is carried out for $\bar{x} = const$ and for the phases $\bar{q}$ which increase linearly in time.

Taking the inequalities (3.17) and (3.16) into account, we obtain the following estimate

$$\int_{t_0}^{t_0+T} \phi(\bar{x}, \bar{q}) \, dt < -\frac{3}{4}\delta T, \qquad (3.18)$$

where $\phi(\bar{x}, \bar{q})$ is represented by the expression (3.10), $\bar{x} = const$ and is located in the annulus $\eta <\| \bar{x} - x_0 \|< \epsilon$. Thus, condition (c) of Theorem 2.1 is fulfilled. The theorem is proved.

The condition of the isolated resonance is an essential characteristic of the nonlinearity of the system. If there are several slow variables, then this condition for systems of the generic form (1.1) is fulfilled in exceptional (degenerate) cases. However, some additional information may considerably simplify the problem. This is the case in the planetary three – body problem which will be considered in Chapter 5, where the integral of the system, i.e. the Hamiltonian will be used.

If there is only one slow variable x in the system (1.1), then this condition is fulfilled for almost all systems.

Condition (3.8) of the theorem states that the non – oscillating part $R_\epsilon(x, r_0, \Omega)$ of the derivative of $v_0(x)$ is small compared to the dissipative part $-\nabla v_0 X_0(x)$ in the neighbourhood of the point x_0. Here, the dependence of the magnitude of R_ϵ on the size of the ϵ-neighbourhood is of considerable importance. The technique for estimating $R_\epsilon(x, r_0, \Omega)$ must take the structure and the convergence rate of the series $X(x, q)$ into consideration. Resonance terms (if any) must also be taken into account, since not all frequencies $k\omega(x)$, which may be small, are represented in the series $X(x, q)$. In what follows we will consider examples of ϵ dependence of R_ϵ.

It is convenient to introduce the quantity $\bar{R}_\epsilon(r_0, \Omega)$ as follows

$$\bar{R}_\epsilon(r_0, \Omega) = \max_{|x-x_0|<\epsilon} \| \nabla v_0 \| R_\epsilon(x, r_0, \Omega). \qquad (3.19)$$

In the case of one variable x it is possible to draw graphs of the functions $\bar{R}_\epsilon$ and $\nabla v_0 X_0(x)$, and thus to illustrate the condition (3.8) of the theorem. We plot $\bar{R}_\epsilon$ as a function of ϵ on the y – axis and $-\frac{1}{2}\nabla v_0 X_0(x)$ as a function of x on the x – axis. To make (3.11) negative, we require that for $\parallel x - x_0 \parallel < \epsilon$ and $\epsilon < \epsilon_0$ the function $-\frac{1}{2}\nabla v_0 X_0(x)$ increases faster than $\bar{R}_\epsilon$ as x moves away from x_0; for any ϵ, $0 < \epsilon < \epsilon_0$ there exists $\gamma > 0$ such that for $\gamma < \parallel x - x_0 \parallel < \epsilon$ the inequality

$$\frac{1}{2}\nabla v_0 X_0(x) + \bar{R}_\epsilon \leq 0$$

holds.

Then, for any narrower region $\gamma < \eta < \parallel x - x_0 \parallel < \epsilon$ there exists $\delta > 0$ such that

$$\nabla v_0 X_0(x) + \bar{R}_\epsilon \leq -\delta. \tag{3.20}$$

For $\eta < \parallel x - x_0 \parallel < \epsilon$, δ can be chosen as

$$\delta = \min \parallel v_0 X_0(x) + \bar{R}_\epsilon \parallel . \tag{3.21}$$

Then, condition (c) of Theorem 2.1 is fulfilled.

2. Now we examine a two – frequency system. In this case the resonance condition

$$k\omega(x) = k_1\omega_1(x) + k_2\omega_2(x) = 0$$

may be conveniently written in the form

$$\frac{k_1}{k_2} = -\frac{\omega_1(x)}{\omega_2(x)} = \kappa(x), \tag{3.22}$$

with restrictions imposed directly on the function $\kappa(x)$. To begin with, we consider the case of a single variable x, where the function $\kappa(x)$ is required to be strictly monotone on the right and on the left of the point x_0.

Let us consider a resonance vector k_0 which satisfies the condition $\omega(x_0)k_0 = 0$. The condition that the function $\kappa(x)$ is strictly monotone means that the resonance is isolated (3.6):

$$\omega(x)k_0 = k_{10}\omega_1(x) + k_{20}\omega_2(x) = k_{10}\omega_2(x)(\kappa(x) - \kappa(x_0)). \tag{3.23}$$

The vectors $\omega(x)$ and $\omega(x_0)$ satisfy the inequality

$$\sin(\omega(x),\omega(x_0)) \geq \frac{\omega_2(x)\omega_2(x_0)}{|\omega(x)|\,|\omega(x_0)|}(\kappa(x) - \kappa(x_0)).$$

Since $\kappa(x)$ is strictly monotone, there exists a monotonically increasing function $\alpha(\epsilon)$, $\alpha(0) = 0$, which may be interpreted as a maximum angle between the vectors $\omega(x)$ and $\omega(x_0)$ for $|x - x_0| < \epsilon$.

The vectors κ which together with the vector $\pm\kappa_0$ form an angle not exceeding $2\alpha(\epsilon)$ will be grouped into the set $\bar{K}_\epsilon(r_0,\Omega)$; we denote such vectors $\bar{\kappa}$. The vectors $\tilde{\kappa} \in \tilde{K}_\epsilon(\Omega)$ form with the vector $+\omega(x)$ the angle not exceeding $\frac{1}{2}\pi - \alpha(\epsilon)$, since $\tilde{\kappa}\omega(x) = |\tilde{\kappa}|\,|\omega(x)|\cos(\tilde{\kappa},\omega)$; to estimate the oscillating frequencies $\tilde{\kappa}\omega(x)$ the function $\Omega(\epsilon)$ may be chosen as follows

$$\Omega(\epsilon) = |\omega(x_0)|\alpha(\epsilon). \tag{3.24}$$

Now let us estimate from below the value of the vectors $\bar{k}$ belonging to $\bar{K}_\epsilon$; for any $\bar{k}$ we have $|[\bar{k}, k_0]| \leq |\bar{k}|\,|k_0|\sin 2\alpha(\epsilon)$, whence

$$|\bar{k}| \geq \frac{|k_1 k_{20} - k_2 k_{10}|}{|k_0|}\frac{1}{\sin 2\alpha(\epsilon)}.$$

Since $\min |k_1 k_{20} - k_2 k_{10}| = 1$, we have

$$|\bar{k}| \geq \frac{1}{|k_0|\sin 2\alpha(\epsilon)}. \tag{3.25}$$

Thus, given the condition that ratios of the frequencies on the right and on the left of x_0 are strictly monotone, the conditions of Theorem 4.2 are satisfied for two frequencies and one slow variable. Therefore, for systems with an asymptotically stable averaged system (i.e when a Lyapunov function $v_0(x)$ exists) the point x_0 is (x, μ) – stable.

Estimates similar to (3.25) will also hold for two – frequency systems in multi – dimensional cases ($s \geq 2$), if the function $\kappa(x) - \kappa(x_0)$ has a strict extremum at the point x_0. It should be noted, however, that the assumption of existence of strict extremum at the point x_0 is rather restrictive.

An equation of the form (3.24) is valid when the gradient of the function $\kappa(x)$ differs from 0 in a neighbourhood of the point x_0. For two variables the gradient has the form

$$|\nabla \kappa(x)| = \frac{1}{\omega_2^2(x)} \sqrt{\begin{vmatrix} \omega_1 & \omega_2 \\ \dfrac{\partial \omega_1}{\partial x_1} & \dfrac{\partial \omega_2}{\partial x_2} \end{vmatrix}^2 + \begin{vmatrix} \omega_1 & \omega_2 \\ \dfrac{\partial \omega_1}{\partial x_2} & \dfrac{\partial \omega_2}{\partial x_2} \end{vmatrix}^2}. \qquad (3.26)$$

Now let us consider a three – frequency system with one slow variable ($m = 3$, $s = 1$). We assume existence of vectors k_r, ($r = 1,, 2, \ldots, \bar{r}$) which are orthogonal to the vector $\omega(x_0)$. Let the plane P be a span of these vectors and be orthogonal to the vector $\omega(x_0)$. We represent the vector $\omega(x)$ as $\omega(x) = \omega(x_0) + \omega'(x)(x - x_0) + o(\| x - x_0 \|)$, assuming the necessary smoothness. To satisfy the condition that the resonance is isolated (3.6), it is sufficient to require that the vector $\omega'(x_0)$ is not orthogonal to any of the vectors k_r, $1 \leq r \leq \bar{r}$, $\omega'(x_0)k_r \neq 0$. In this case the estimate

$$|k_r \omega(x)| > \rho_1 \cdot \text{const}, \quad \text{if} \quad \| x - x_0 \| > \rho_1 \qquad (3.27)$$

holds for the combinative resonance frequency $\lambda_{k_r}(x) = k_r \omega(x)$. In the case when $\omega(x_0)k_r = 0$ and $\omega'(x_0)k_r = 0$, $1 \leq r \leq \bar{r}$, but

$$\Delta(\omega_1, \omega_2, \omega_3)|_{x=x_0} = \begin{vmatrix} \omega_1 & \omega_2 & \omega_3 \\ \omega_1' & \omega_2' & \omega_3' \\ \omega_1'' & \omega_2'' & \omega_3'' \end{vmatrix}_{x=x_0} \neq 0, \qquad (3.28)$$

the estimate for the frequencies is of the second order of smallness

$$|k_r \omega(x)| > \rho_1^2 \cdot \text{const}, \quad \text{if} \quad \| x - x_0 \| > \rho_1 \qquad (3.29)$$

To construct the function $\Omega(\epsilon)$ and to divide the vectors k into the classes $\tilde{K}_\epsilon(\Omega)$ and $\bar{K}_\epsilon(\bar{r}, \Omega)$, we require that the vector $\omega'(x_0)$ form the angle α_0 with the plane P,

$$\alpha_0 < \frac{1}{2}\pi. \qquad (3.30)$$

The angle β between the vectors $\omega(x)$ and $\omega(x_0)$ can be represented as

$$\beta(x) = \frac{|\omega'(x_0)|}{|\omega(x_0)|} \parallel x - x_0 \parallel \cos \alpha_0,$$

$$(3.31)$$

$$\beta(\epsilon) = \epsilon \cos \alpha_0 \frac{|\omega'(x_0)|}{|\omega(x_0)|},$$

$$(3.32)$$

if $\parallel x - x_0 \parallel < \epsilon < \epsilon_0$, with sufficiently small ϵ_0.

Now, we include to the class $\bar{K}_\epsilon(\bar{r}, \Omega)$ from the vectors $\bar{k}$ which form an angle not exceeding $2\beta(\epsilon)$ with the plane P; the class $\tilde{K}_\epsilon(\Omega)$ consists of all vectors $\bar{k}$ and k_r. The angle between vectors $\bar{k}$ and the vector $\omega(x)$ does not exceed $\frac{1}{2}\pi - \beta(\epsilon)$, thus $\beta(\epsilon)$ can be chosen to be the function $\Omega(\epsilon)$ in (3.7).

The properties of frequencies, expressed in the condition of isolation of resonances make it possible to associate the main resonance harmonics with the oscillatory ones, and thus to minimize restrictions on the dissipative properties of the system. It is obvious that in the systems with considerable dissipation, stability can be only achieved by taking into account properties of the averaged system, and that for such systems it is not necessary to satisfy the condition of isolated resonance.

We would like to point out that isolated resonance is not the only condition that weakens the requirements on the averaged system. Below, we present an example of similar condition of another type related to the properties of coefficients of the Fourier series of $X(x, q)$.

Let us consider the system (1.1), assuming, as above, that the point x_0 is an asymptotically stable equilibrium position of the averaged system (1.2) and that in the neighbourhood of the point x_0 there exists a positive – definite Lyapunov function $v_0(x)$ with negative – definite derivative. We assume that several combinative frequencies $\lambda_k(x) = k\omega(x)$ become zero on curves (surfaces) passing through the point x_0. The corresponding phases are denoted by $Q = kq$. The sum of the corresponding terms on the RHS of the system (1.1) is denoted by $\bar{X}(x, Q)$. Thus $\bar{X}(x, Q)$ contains the terms which do not oscillate on the resonance curves passing through the point x_0. We impose restrictions on coefficients of the series appearing on RHS: let the resonance terms united in the sum $\bar{X}(x, Q)$ have higher order of smallness with respect to $\| x - x_0 \|$ than $X_0(x)$, to wit

$$|\bar{X}(x, Q)| \leq \| x - x_0 \|^{\alpha} \| X_0(x) \|, \quad \alpha > \alpha_0 > 0. \tag{3.33}$$

It would be interesting to take more precise account of properties of the coefficients in the neighbourhood of the equilibrium position x_0 of the averaged system, especially when the coefficients and the RHS of the averaged system have the same order of smallness at x_0.

If a resonance curve (surface)

$$k_0 \omega(x) = 0 \tag{3.34}$$

coincides with the curves $X_0(x) = 0$ or if $X_0(x) \equiv 0$, then one clearly should investigate not the stability of a point but of this curve. It is appropriate to replace the coordinates used previously with new coordinates related to the resonance curve; one of the coordinates should be directed along a vector normal to the curve, and investigation of stability should be performed along this coordinate.

4.4 Stability of Multi – Frequency Systems on Finite Time Interval

In Theorem 4.2, the stability of a resonance point in the multi – frequency system (1.1) was achieved at the expense of the asymptotic stability of the averaged system (1.2) at this point.

Now, we consider the case where the resonance point provides a stable equilibrium position in the averaged system, i.e. when there is no asymptotic stability. Under such minimal assumptions we determine the length of the time interval T during which the solution of the system

(1.1) does not leave the ϵ-neighbourhood (in the variables x) of the resonance point being investigated for stability. The length of the time interval T depends on the value of μ and T infinitely increases as $\mu \to 0$.

In this case, investigation of stability is carried out by applying Theorem 2.3.

Concerning to the averaged system (1.2), we assume that the point x_0 is a stationary stable point. Stability is ensured by existence of the positive – definite Lyapunov function $v_0(x)$ whose derivative is non – positive: $\dot{v}_0 = \mu \nabla v_0 X_0(x) \leq 0$. In this way we satisfy conditions (a) and (b) of Theorem 2.3.

Having set $\epsilon > 0$ $(\epsilon < H)$, we choose a number w_0 such that the surface

$$v_0(x) = w_0 \tag{4.1}$$

lies entirely in the ϵ-neighbourhood. Let us choose a number $\sigma > 0$ such that it obeys the condition $2\sigma < w_0$, and let us consider the surface

$$v_0(x) = w_0 - 2\sigma. \tag{4.2}$$

Since the function $v_0(x)$ is positive – definite, there exists $\eta > 0$ such that the η-neighbourhood lies inside the surface (4.2).

We assume that at the point x_0 the condition for isolated resonance (see Sect. 3) is satisfied.

Having set $\epsilon > 0$, as in Sect. 3, we split the frequencies $k\omega(x)$ into two classes of resonance and oscillating frequencies with the estimates from below (3.6) and (3.7). For the classes of vectors k and the sums of harmonics we will retain notations of Sect. 3.

Let us consider a retarded Lyapunov function

$$v = v_0(x) + \mu U(x, q). \tag{4.3}$$

The function $U(x, q)$ is defined as a solution of the equation

$$\frac{\partial U}{\partial q}\omega(x) = -\frac{\partial v_0}{\partial x}\tilde{X}(x, q), \tag{4.4}$$

bounded in the annulus $\eta <\| x - x_0 \|< \epsilon$.

Such a solution exists since the estimates (3.6) and (3.7) are valid for the combinative frequencies $k\omega(x)$ corresponding to $\tilde{X}(x,q)$. This ensures the fulfilment of condition (c) of Theorem 2.3.

Let us introduce the notation

$$\bar{R}_\epsilon(r_0, \Omega(\epsilon)) = \max_{\|x - x_0\| \le \epsilon} |\nabla v_0| R_\epsilon(x, r_0, \Omega). \qquad (4.5)$$

Having differentiated the perturbed Lyapunov function (4.3) by virtue of the system (1.1) and having taken (4.4) into account, we get

$$\frac{dv}{dt} = \mu \frac{\partial v_0}{\partial x} X_0(x) + \mu \left[\frac{\partial v_0}{\partial x} R_\epsilon + \mu \frac{\partial U}{\partial x} X(x,q) + \mu \frac{\partial U}{\partial q} \Phi \right] = \qquad (4.6)$$

$$\mu \nabla v_0 X_0(x) + \mu \phi(x, q, \epsilon, \mu).$$

Hence, in the annular region

$$|\phi(x, q, \epsilon, \mu)| \le \phi_0 = \bar{R}_\epsilon(r_0, \Omega(\epsilon)) + O(\mu). \qquad (4.7)$$

All the conditions of Theorem 2.3 are now fulfilled. Therefore, in the variables x, all the solutions emanating from the η-neighbourhood of the point x_0, for the time not exceeding

$$T(\epsilon, \mu) = \frac{\sigma(\epsilon)}{2\phi_0 \mu} \qquad (4.8)$$

remain in the ϵ-neighbourhood of the resonance point x_0 for $0 < \mu < \mu_0$. The value $\mu_0(\epsilon)$ depends on $\sigma(\epsilon)$: in course of proving Theorem 2.3 we assumed that in the annulus $\eta <\| x - x_0 \|< \epsilon$ the condition

$$\mu|U(x,q)| < \frac{\sigma(\epsilon)}{4} \qquad (4.9)$$

is imposed.

The result obtained above can be formulated in a form of the following theorem

Theorem 4.3 *Suppose that:*

a) The averaged system (1.2) *has a positive - definite Lyapunov function* $v_0(x)$ *whose derivative is non - positive* $\nabla v_0 X_0(x) \leq 0$.

b) The resonance at the point x_0 *is* r_0 - *isolated for some* r_0.

c) For any $\epsilon > 0$ *(* $\epsilon < \epsilon_0$ *) there exists a number* r_0 *and a function* $\Omega(\epsilon)$ *and, hence, a division of vectors* $k \in K$ *into subsets* $K(r_0)$, $\tilde{K}_\epsilon(r_0, \Omega)$, $\bar{K}_\epsilon(r_0, \Omega)$ *such that*

$$\max_{|x-x_0|\leq\epsilon} |\nabla v_0| R_\epsilon(x, r_0, \Omega) = \bar{R}_\epsilon(r_0, \Omega) \to 0 \quad for \quad \epsilon \to 0.$$

d) The functions $X(x, q)$ *and* $\Phi(x, q)$ *are bounded.*

Then for each $\epsilon > 0$ *(* $\epsilon < \epsilon_0 < H$ *) there exist* $\eta(\epsilon)$ *and* $\mu_0(\epsilon)$ *such that for any* μ, $0 < \mu < \mu_0(\epsilon)$, *the solution* $x = x(t)$, $q = q(t)$ *with the initial conditions* $x(0) = \bar{x}_0$, $q(0) = q_0$ *and satisfying in* x *the inequality* $\| \bar{x}_0 - x_0 \| < \eta(\epsilon)$, *fulfils the inequality* $\| x(t) - x_0 \| < \epsilon$ *for all* t *not exceeding* $T(\epsilon, \mu) = \sigma(\epsilon)[2\phi_0\mu]^{-1}$.

4.5 Stability of Multi – Frequency Problems of Nonlinear Mechanics

1. We consider a general second order equation which describes oscillations of nonlinear system affected by a small force, nonlinear with respect to the coordinates and velocities and periodically dependent on several phases. We also examine tho particular problems which can be reduced to the above – mentioned equation: the problem of plane oscillation of pendulum whose suspension point makes small harmonic oscillations along two orthogonal directions in the oscillation plane, with frequencies commensurable with the eigenfrequency of the pendulum (here the stability of the resonance modes will be investigated), and the problem of stability of two – frequency system which describes oscillations of a satellite with respect to the centre of mass. In estimating small denominators, the system is not reduced to one with smaller

number of frequencies. These two problems were considered previously under additional assumptions decreasing the number of frequencies in the system ([11], [17], [10]).

Let us consider the equation

$$\ddot{x} + \omega^2 \sin x = \mu f(x, \dot{x}, \omega_1 t, \dots, \omega_n t), \quad 0 < \mu \ll 1. \tag{5.1}$$

Nonlinear and linearized equations of the form (5.1) were studied by many authors. Detailed studies were undertaken concerning the effect of external periodic forces on oscillating systems which, on an asymptotically large time interval of the length $O(\mu^{-1})$, are close to linear. The stability of resonance amplitudes under the parametric resonance in the single – frequency case was considered in [11] and [87].

Eq. (5.1) and the problems related to it are investigated under assumption that the frequencies of oscillations of the system itself and the frequencies of the external forces are comparable in magnitude. In this case, the problem is reduced to a multi – frequency system of first order differential equations in the standard form. The small denominators appearing here are typical to multi – frequency systems, and they are estimated by means of the size of the neighbourhood of the point whose stability is being investigated [9].

Eq. (5.1) with $\mu = 0$ is integrable by means of elliptic functions and its general solution describes the oscillations

$$\alpha = 2\arcsin(k\operatorname{sn}(2\pi^{-1}K(k)\phi, k)) \tag{5.2}$$

or the rotations

$$a = 2\operatorname{am}(\pi^{-1}K(k)\phi, k). \tag{5.3}$$

Here, $\phi = \omega_0(k)(t + t_0)$, $t_0 =$ const, $K(k)$ is the total elliptic integral of first kind, $\omega_0(k)$ is the frequency of motion of the system itself, with

$$\omega_0(k) = \frac{\pi\omega}{2K(k)} \quad \text{in the case of oscillations}$$

or

$$\omega_0(k) = \frac{\pi\omega}{kK(k)} \quad \text{in the case of rotations.} \tag{5.4}$$

Let us investigate the case of oscillations (rotations can be investigated in a similar manner). Since $\alpha_{max} = 2\arcsin(k)$, it is convenient to select k as the variable describing the amplitude of oscillations, while the phase is described by the angle ϕ.

Let us transform Eq. (5.1) to a multi – frequency system in the standard form with respect to the variables k and ϕ. To this end, we assume that $f(x, y, \bar{z}) = f(x, y, z_1, \ldots, z_n)$ is a 2π – periodic function in the variables $z_1, \ldots, z_n$ and is $[\frac{n}{2}] + 2$ times continuously differentiable and satisfies the equation

$$f_0(x, y) = (2\pi)^{-n} \int_0^{2\pi} f(x, y, \bar{z}) \, d\bar{z} = 0. \tag{5.5}$$

In Eq. (5.1) we perform the change of variables

$$x = \alpha(k, \phi), \quad \dot{x} = \omega_0(k)\alpha'_\phi(k, \phi) \tag{5.6}$$

where $\alpha'_\phi(k, \phi) = \dfrac{\partial \alpha}{\partial \phi}(k, \phi)$ and introduce the action integral

$$I(k) = \frac{\omega_0(k)}{2\pi} \int_0^{2\pi} \alpha_\phi^2(k, \phi) \, d\phi, \quad I'(k) = D(k). \tag{5.7}$$

For the new variables k and ϕ we obtain the multi – frequency system of standard – form differential equations

$$\dot{k} = \mu f(\alpha(k, \phi), \omega_0(k)\alpha'_\phi(k, \phi), \phi_1, \ldots, \phi_n)\frac{\alpha'_\phi(k, \phi)}{D(k)} = \mu F(k, \bar{\phi}),$$

$$\dot{\phi} = \omega_0(k) - \mu f(\alpha(k, \phi), \omega_0(k)\alpha'_\phi(k, \phi), \phi_1, \ldots, \phi_n)\frac{\alpha'_\phi(k, \phi)}{D(k)} =$$

$$= \omega_0(k) + \mu\Phi(k\bar{\phi}), \tag{5.8}$$

$$\dot{\phi}_i = \omega_i, \quad \phi_i = \omega_i t, \quad i = 1, \ldots, n,$$

$$\bar{\phi} = (\phi_1, \ldots, \phi_n).$$

The frequency vector $\Omega = (\omega_0(k), \omega_1, \ldots, \omega_n)$ consists of the angular frequency $\omega_0(k)$ and the frequencies of the perturbing force.

The quantity k_0 will be called the resonance point of the system if there exists an integer – valued vector $\vec{p} = (p_0, p_1, \ldots, p_n)$ such that

$$p_0 \omega_0(k) + p_1 \omega_1 + \cdots + p_n \omega_n) = 0. \tag{5.9}$$

In order to investigate the Lyapunov stability of the resonance amplitude k_0, we apply the generalized Lyapunov second method.

Let us take $V_0 = |k - k_0|$ as a non – perturbed Lyapunov function and consider segments $\eta < |k - k_0| < \epsilon$, where $0 < \epsilon < \epsilon_0$ and ϵ_0 (by virtue of the properties of the function $\omega_0(k)$) does not exceed the distance from k_0 to the nearest resonance amplitude. Then the ϵ-neighbourhood of the point does not contain any other resonance values for $0 < \epsilon < \epsilon_0$.

For $\eta < |k - k_0| < \epsilon$ we define the perturbed Lyapunov function as follows

$$V(k, \bar{\phi}) = V_0(k) + \mu V_1(k, \bar{\phi}),$$

$$\tag{5.10}$$

$$V_1(k, \bar{\phi}) = \sum_{m \neq 0} \imath \frac{F_m(k) \operatorname{sgn}(k - k_0)}{(\Omega, m)} e^{\imath(m, \phi)},$$

where $F_m(k)$ are the coefficients of the Fourier series for $F(ik, \bar{\phi})$.

In the annulus defined above, the function $V_1(k, \bar{\phi})$ exists and is bounded, since by virtue of assumptions made above, the function $F(k, \bar{\phi})$ is expanded into uniformly and absolutely convergent Fourier series with $F_0(k) = 0$; for the small denominator (Ω, m) the estimate

$$m_0 \omega_0(k) + \sum_1^n m_i \omega_i = m_0(\omega_0(k) - \omega_0(k_0)) \approx$$

$$\tag{5.11}$$

$$\approx \omega_0'(k_0)|k - k_0| = O(\epsilon, \eta)\omega_0'(k_0), \quad \omega_0'(k_0) < 0,$$

holds by virtue of the properties of the function $\omega_0(k)$.

Let us differentiate the function (5.10) along the solution of the system (5.8). We obtain

$$\frac{dV(k, \bar{\phi})}{dt} = O(\mu^2). \tag{5.12}$$

Thus, all conditions of Theorem 4.3 are fulfilled for the system (5.8) and for the Lyapunov function (5.10), and the estimate $T = O(\mu^{-2})$ is valid for the time interval during which the solution $k(t)$ remains in the ϵ-neighbourhood of the point k_0.

All the estimates above are valid for the first resonances until the difference between two subsequent resonances becomes a quantity of order μ.

2. Let us apply the results obtained above to investigate stability of the resonance amplitudes of pendulum whose suspension point performs planar oscillation according to the lw

$$x = a \cos \nu_1 t, \quad y = \beta \sin(\nu_2 t + \chi) \tag{5.13}$$

where a, b, χ are constants.

Let ψ be the angle of deviation from the lower equilibrium position, $\omega = \sqrt{\frac{g}{l}}$ be the eigenfrequency of pendulum oscillations, where l is the length of the pendulum. Suppose that $\frac{a}{l} \ll 1$. This means that the suspension point performs small oscillations. Equations of motion of the pendulum have the following form

$$\ddot{\psi} + \sin \psi = \mu(\cos \phi_1 \cos \psi + \delta \sin \phi_2 \sin \psi),$$

$$\omega_1 = \frac{\nu_1}{\omega}, \quad \omega_2 = \frac{\nu_2}{\omega}, \quad \phi_1 = \nu_1 t, \quad \phi_2 = \nu_2 t + \chi, \tag{5.14}$$

$$\delta = \frac{b}{a}\left(\frac{\omega_2}{\omega_1}\right)^2, \quad \tau = \omega t, \quad \mu = \frac{a}{l}.$$

The change of variables (5.6) results in a three – frequency standard – form system

$$\dot{k} = \mu \left[\frac{1}{2} \mathrm{cn} \frac{2K(k)}{\pi} \phi \left(1 - 2k^2 \mathrm{cn}^2 \frac{2K(k)}{\pi} \phi \right) \cos \phi_1 + \right.$$

$$\left. + k \mathrm{sn} \frac{2K(k)}{\pi} \phi \, \mathrm{cn} \frac{2K(k)}{\pi} \phi \, \mathrm{dn} \frac{2K(k)}{\pi} \phi \sin \phi_2 \right], \tag{5.15}$$

$$\dot{\phi} = \frac{\pi}{2K(k)} + \mu \Phi(k, \bar{\phi}), \quad \dot{\phi}_1 = \omega_1, \quad \dot{\phi}_2 = \omega_2.$$

Let us expand the RHS of the first equation from (5.15) into the Fourier series (it should be noted that in this case the function $F(k, \bar{\phi})$ is analytic in $\bar{\phi}$):

$$F(k, \bar{\phi}) = \sum_n b_n(k)[\cos(n\phi + \phi_1) + \cos(n\phi - \phi_1)] +$$

$$+ \sum_m c_m(k)[\cos(m\phi - \phi_2) - \cos(m\phi + \phi_2)] +$$

$$+ \sum_m a_m(k)[\sin(m\phi + \phi_2) + \sin(n\phi - \phi_2)], \qquad (5.16)$$

$$\int_0^{2\pi} F(k, \bar{\phi}) \, d\bar{\phi} = 0.$$

The coefficients $a_m(k)$, $b_n(k)$, $c_m(k)$, tends exponentially to zero as $m, n \to \infty$. Therefore, if

$$\omega_0(k_0) = \frac{\omega_1}{n}, \qquad \omega_0(k_0) = \frac{\omega_2}{m} \qquad (5.17)$$

then the non – oscillating terms of the form $const \cdot \cos \tau_0$, $const \cdot \sin(\tau_0 - \chi)$, $const \cdot \cos(\tau_0 - \chi)$ will appear on the RHS of the first equation of (5.15), i.e. the resonance phenomena will be observed. In accordance with the above result, the resonance point k_0, which satisfies one of the equations (5.17) will be stable during the time $T = O(\mu^{-2})$.

3. We investigate stability of resonance modes in the problem of oscillations and rotations of a satellite moving along the elliptic orbit in the central gravitational field. A system describing the satellite motion is the two – frequency one. The frequency vector consists of the frequency of the satellite rotation around the mass centre and the frequency of its rotation around the Earth.

The resonance oscillations and rotations of a satellite around the mass centre (in the orbital plane) were studied before ([17], [10], [79]) by the Krylov – Bogolyubov averaging method ([11], [87]) under stringent restrictions on the resonance frequencies, restrictions that made it possible to reduce the initial two – frequency system to a single – frequency one; small oscillations of a satellite and resonance phenomena in the motion of the Moon were also studied.

Let the main axis of inertia of the satellite be perpendicular to the orbital plane. Let B be the moment of inertia relative to the main axis, A and C $(A \geq C)$ be the moments of inertia relative to the remaining two axes. Then, up to the quantity of the magnitude of order of the ratio of the satellite size and the size of the orbit, the equation of motion of the satellite has the form

$$(1 + e \cos \theta) \frac{d^2 \delta}{d\theta^2} - 2e \sin \theta \frac{d\delta}{d\theta} + 3a^2 \sin \delta = 4e \sin \theta. \qquad (5.18)$$

Here, $a^2 = \frac{A-C}{B} \leq 1$, δ is the doubled angle between the radius vector of the centre of mass and the axis of inertia (with respect to which the moment of inertia is C), e is the eccentricity of the orbit, θ is the angular distance of the radius vector from the perigee of the orbit.

At $e = 0$, Eq. (5.18) is reduced to the equation of a pendulum and describes motion of the satellite along a circular orbit.

Let us examine the resonance case of $e \ll 1$ with the orbit close to the circular one. Up to the quantities of order of $O(e^2)$ we have

$$\frac{d^2 \delta}{d\theta} + 3a^2 \sin \delta = ef\left(\theta, \delta, \frac{d\delta}{d\theta}\right), \qquad (5.19)$$

where

$$f\left(\theta, \delta, \frac{d\delta}{d\theta}\right) = 4 \sin \theta + 2 \sin \theta \frac{d\delta}{d\theta} + 3a^2 \cos \theta \sin \delta.$$

The change of the variables

$$\delta = \alpha(k, \phi), \quad \phi = \omega_0(k)(\theta + \theta_0), \quad \theta_0 = \text{const},$$

similar to (5.6), transform (5.19) to the standard form

$$\dot{k} = eF(k, \phi, \theta),$$
$$\dot{\phi} = \omega_0(k) + e\Phi(k, \phi, \theta), \qquad (5.20)$$
$$\dot{\theta} = 0,$$

where

$$F(k,\phi,\theta) = f(\theta,\alpha(k,\phi),\omega_0(k)\alpha'_\phi(k,\phi))\frac{\alpha'_\phi(k,\phi)}{D(k)},$$

$$\Phi(k,\phi,\theta) = -f(\theta,\alpha(k,\phi),\omega_0(k)\alpha'_\phi(k,\phi))\frac{\alpha'_\phi(k,\phi)}{D(k)}$$

Let us expand the RHS of this system into Fourier series and use elementary trigonometry. We obtain terms of the form $G(k,\theta)$, $g_n(k)\sin(\theta + n\phi)$, $g'_n(k)\sin(\theta - n\phi)$. It should be noted that for $\omega_0(k) = \dfrac{1}{n}$

$$g'_n(k)\sin(\theta - n\phi) = g'_n(k),$$

$$\sin(\theta - n\omega_0(k)(\theta + \theta_0)) = -g'_n(k)\sin\theta_0.$$

Thus, at $\omega_0(k) = \dfrac{1}{n}$, where n is an integer, slowly varying terms of the form $-g'_n(k)\sin\theta_0$ will appear on the RHS of the system, i.e. resonance phenomena will appear both in the case of oscillations and rotations. According to results obtained above, resonances of the form (5.21) will be stable in the slow variable k during the time period $T = O(e^{-2})$.

CHAPTER 5

Stability of Orbits in Three – Body Problem

In this chapter we consider stability in the point model of the three – body problem, and we suggest a model based on the assumption that planets are oblate rotating bodies whose axes of rotation are inclined to the orbital plane at constant angle. The latter model will be called hydrodynamic.

In the point model we obtain stability of the system resonance point on the asymptotically large interval $(T \sim \mu^{-2}, \mu^{-3}$) under real initial deviations and values of the small parameter μ equal to $10^{-3} - 10^{-4}$. The time unit is taken here to be equal to the rotation period of the planet. Use is made of the fact that the system is Hamiltonian, the canonical system of Poincaré is chosen, the two – frequency problem is reduced to a single – frequency problem, using the known integrals of motion. The latter enable us to estimate small denominators or to use the isolated character of the resonance to be investigated for stability. The generalized (perturbed) Lyapunov function is constructed.

140

In the hydrodynamic model, we construct an additional term in the Hamiltonian, the term which is determined by the oblate shape of the planet and by the inclination of its axis of rotations. The system of canonical variables, the condition that the resonance is isolated, and reduction to one – dimensional problem in slow variables is done exactly as in the point model. The main difficulty is caused by the fact that we must construct the perturbed Lyapunov function as a polynomial in powers of μ. Since the mean is negative, the system is stable on the infinite interval.

5.1 Orbit Stability in Three – Body Problem and Description of the Models

In the classical three – body problem, one studies the motion of two planets in the gravitational field of a third, more massive body. The main question here is that of stability, i.e. conservation of the semi – axes of the planet in proximity of their initial values and of preservation of smallness of the eccentricities and inclinations over the infinite time interval. Here, the angular variables can change in an arbitrary fashion. The first fundamental result in solving this problem was obtained in Laplace's Treatise ([1]). After that, the problem was investigated by many authors ([11], [96], [95], [65], [8], [38], [39], [86]). These authors dealt with the point model without dissipation, so they could detect no stability on the infinite time interval. Among the publications devoted to the three – body problem, one should mention the works by H. Poincaré ([1], [96]), where the problem is investigated with the help of methods of perturbation theory. These works helped us to choose the most convenient (from the point of view of stability investigations) canonical coordinate system and the system of integrals of motion.

Among more recent works we would like to mention Refs. [38], [39], [86], where stability in the three – body problem was investigated according to the Lagrange method.

Recently, papers have appeared (e.g. [107]) which take account of size of a planet, with assumptions that planets are solid bodies. Equations for the translational – rotational movement were formulated for such models and the corresponding integrals were found.

In Ref. [59] another model for a planetary system was suggested: planets were assumed to be liquid rotating bodies with a noticeable inclination angle β between the axis of rotation and the Laplace static plane, and with a noticeable coefficient α of elongation along the equator $\alpha = (r_e - r_p)/r_e$, where r_e and r_p are the equatorial and polar radii, respectively.

It will suffice if at least one of the planet possesses these properties. A real three – body system (Sun – Jupiter – Saturn) is characterized by several small parameters. They are μ , the ratio of planet mass M_p to solar mass M_S , being approximately 10^{-3}; $\bar{\mu}$, the ratio of the radius of the planet to its distance from the Sun, $\bar{\mu} \simeq 10^{-4}$. It is clear that for Jupiter and Saturn these parameters are of the same order of smallness. The coefficient of elongation over the equator (oblateness) and the inclination angles of the axes of rotation are of the orders $\alpha_U \simeq \alpha_{Sat} \simeq 0.1$, $\beta_U \simeq \beta_{Sat} \simeq 30°$. Further, in description of the model of the three – body problem, we assume $0 < \mu \ll 1$, $0 < \bar{\mu} \ll 1$, $\alpha < 1$, $\sin\beta < 1$.

Elongation along the equator (oblateness) of a planet, combined with a noticeable inclination of the axes of rotation results in small periodic dependence of the attracting force on the angle ($\bar{\mu}^2$ being the parameter of smallness) that determines position of the planet on the orbit. The purpose of the investigation is to discover a stabilizing combination of factors under assumption that the coefficient of elongation α and the spatial location of the axis of rotation remain unchanged in time. Precession, which is unavoidable in the solid body model is absent in our case: it is neutralized by hydrodynamical current which provides the model with its dissipative character. This is the reason why we called the model hydrodynamic. The results obtained in the framework of this model hold during the time when the model is accurate. In the course of the evolution, which is affected not only by mechanical factors, the stabilizing effect of the parameters α and β will become weaker as these parameters decrease.

We investigate stability in the above – stated problems with the help of the generalized Lyapunov method. The three – body problem is a two – frequency one. The real anomalies of the two planet are the rapidly varying phases. When investigating the point model, we will make use of Theorem 4.3; Theorem 4.1 will be applied to the hydrodynamic model.

5.2 Canonical Variable, Equations and Integrals of Motion in the Point-like Three – Body Problem

In order to investigate stability in the three – body problem, it is convenient ([96], [95]) to choose the canonical variables L, λ, σ, ω. The variables L and σ are expressed in terms of the orbital parameters and masses of planets:

$$L = m\sqrt{M}\sqrt{a}$$

where a is the longer semi-axis, m and M are the masses of both planets and the Sun, respectively,

$$\sigma_1 = L(1 - \sqrt{1 - e^2}),$$

e being the eccentricity. For the planets of the Solar system $e \ll 1$, therefore, $\sigma_1 \simeq e^2$;

$$\sigma_2 = (L - \sigma_1)(1 - \cos i),$$

σ_2 being the angle of inclination. For the planets in the Solar system $i \ll 1$, therefore $\sigma_2 \simeq i^2$.

As the slow and fast variable we choose: λ, the mean longitude; ω_1, the perihelion longitude; ω_2, the node longitude.

The variables L_1, λ_1, σ_1, ω_1, σ_2, ω_2 describe the first planet, while the variable L_2, λ_2, σ_3, ω_3, σ_4, ω_4 describe the second. The Hamiltonian of the three – body problem in these variables takes the form

$$F = F_0 + \mu F_1 \tag{2.1}$$

where

$$F_0 = -\frac{M_1}{2L_1^2} - \frac{M_2}{2L_2^2} \tag{2.2}$$

Here M_1 and M_1 are constants, μF_1 is the perturbating function, and μ is a small parameter equal to the ratio of masses of the planet to the mass of the Sun (the planet masses are assumed to be of the same order of magnitude).

Then,

$$\mu F_1 = \sum A\sigma_1^{q_1}\sigma_2^{q_2}\sigma_3^{q_3}\sigma_4^{q_4} \cos(\sum k_i\lambda_i + \sum p_i\omega_i). \tag{2.3}$$

Here, k_i and p_i are integers over which summation is made; the coefficient A depends on L only; q_i are positive integer, $2q_i \geq |p_i|$.

Since dimensions of the terms of the series (2.3) are the same, the order of magnitude of the expansion terms of the perturbating function is equal to powers of the eccentricities and inclination. We note that properties of the perturbating function were studied in detail by H. Poincaré ([96], [95]).

Using the Hamiltonian (2.1), we find the equations of motion

$$\frac{d\lambda}{dt} = \frac{\partial F}{\partial L}, \quad \frac{dL}{dt} = -\frac{\partial F}{\partial \lambda},$$

$$\frac{d\omega}{dt} = \frac{\partial F}{\partial \sigma}, \quad \frac{d\sigma}{dt} = -\frac{\partial F}{\partial \omega}.$$

By μh_{kp} we denote the coefficients of the series (2.3); the argument of the cosine will be denoted by θ ; we also introduce the frequencies

$$n_1(L) = \frac{M_1}{L_1^3}, \quad n_2(L) = \frac{M_2}{L_2^3}, \tag{2.4}$$

assuming in what follows that $n_1 \neq n_2$.

In these notations Eq. (1.4) will take the form

$$\frac{d\lambda}{dt} = n + \mu \sum_{k,p} \frac{\partial h_{k,p}}{\partial L} \cos \theta, \quad \frac{dL}{dt} = \mu \sum_{k,p} h_{k,p} k \sin \theta,$$

$$\frac{d\omega}{dt} = \mu \sum_{k,p} \frac{\partial h_{k,p}}{\partial \sigma} \cos \theta, \quad \frac{d\sigma}{dt} = \mu \sum_{k,p} h_{k,p} p \sin \theta.$$

The variables L, σ, ω in Eq. (2.6) vary slowly, and their derivatives are proportional to the small parameter μ. The phases λ_1 and λ_2 vary rapidly (with the frequencies $n + 0(\mu)$). If the combinative frequencies $k_1 n_1(L) + k_2 n_2(L)$ become zero, the slowly varying terms appear on the RHS of Eq. (2.6), i.e. we have a resonance.

Eqs. (2.6) describing the three – body problem have integrals of motion ([96], [95]).

$$\sum L - \sum \sigma = \text{const}, \qquad (2.5)$$

$$\sigma_2(L_1 - \sigma_1 - \frac{1}{2}\sigma_2) = \sigma_4(L_2 - \sigma_3 - \frac{1}{2}\sigma_4). \qquad (2.6)$$

These integral are called area integrals.

The integral (2.7) relates evolution of the variables L and σ, and makes it possible to estimate the evolution of σ if the variables L in the course of their evolution remain close to some point L_0 . Let us represent the integral (2.7) as follows:

$$L_1 + L_2 - (\sigma_1 + \sigma_2 + \sigma_3 + \sigma_4) =$$
$$= L_{1\,in} + L_{2\,in} - (\sigma_{1\,in} + \sigma_{2\,in} + \sigma_{3\,in} + \sigma_{4\,in}),$$

where the subscript *in* denotes the initial values of the variable. Hence, we obtain

$$\Delta L_1 + \Delta L_2 = \Delta \sigma_1 + \Delta \sigma_2 + \Delta \sigma_3 + \Delta \sigma_4. \qquad (2.7)$$

The series (1.3), representing expansion of the perturbation function and the series on the RHS of Eq. (2.6) converge rather rapidly when the eccentricities and inclinations are small since the terms in these series

are proportional to powers of these quantities. Now we show that if
the initial values of these quantities are small, then they remain small,
provided that the variables L remain close to their initial values. From
equation (2.9), we obtain

$$\Delta\sigma_k = \Delta L_1 + \Delta L_2 - \sum_{i \neq k} \Delta\sigma_i$$

and if $\Delta L = O(\epsilon)$, then

$$\Delta\sigma_k = O(\epsilon) - \sum_{i \neq k} \Delta\sigma_i.$$

It should be noted that the quantities σ are proportional to squares
of the eccentricities and inclinations and therefore are positive. Ac-
cording to Eq. (2.10) the $\Delta\sigma_k$ changes most rapidly if $\Delta\sigma_i$ $(i \neq k)$ are
negative, i.e. when σ_i $(i \neq k)$ decrease while $\Delta\sigma_k$ increases. Therefore,
the corresponding dimensionless quantity (eccentricity or inclination)
remains small when ϵ is sufficiently small. Thus, it is not necessary to
investigate the stability of all the variables L and σ : if the values of L
do not leave the ϵ-neighbourhood of some point L_0, then the values of
σ will remain small, provided that ϵ is sufficiently small. This is exactly
as stated in the Laplace theorem ([1]). It follows that the requirement
that the series (2.6) converge rapidly results in certain restriction on
upper bound of ϵ.

The integral (2.8) links the variations in the inclinations of two
planets and it permits a somewhat improved estimate (2.10) for σ_2 and
σ_4.

Let us note that for the planar three - body problem, when the
inclinations are zero, we have

$$\Delta\sigma_1 = O(\epsilon) - \Delta\sigma_3. \tag{2.8}$$

5.3 Resonance Curves and Choice of New Variables

Let us consider the combinative frequencies

$$k_1 n_1(L) + k_2 n_2(L) = \Lambda(L). \tag{3.1}$$

It is important that the main part of the Hamiltonian function (2.2) and, therefore, the frequencies n_1 and n_2, depend only on the variable L, i.e. the longer semi-axes.

The resonance lines determined by the equation

$$k_1 \frac{M_1}{L_1^3} + k_2 \frac{M_2}{L_2^3} = 0 \tag{3.2}$$

are rays emanating from the origin in the L plane:

$$\frac{L_1}{L_2} = \sqrt[3]{\frac{M_1}{M_2}} \sqrt[3]{-\frac{k_2}{k_1}}. \tag{3.3}$$

We will investigate stability of the point L_0 lying on a resonance ray belonging to the vector $k_0 = \{k_{10}, k_{20}\}$.

The curve

$$F_0(L) = F_0(L_0) \tag{3.4}$$

also passes through the point L_0.

Since the function $F = F_0 + \mu F_1$ is an integral of motion and since $\nabla F_0(L) \neq 0$, the integral curve on the plane L remains in the μ-neighbourhood of the curve (3.4), the motion of the integral curve in the direction of the normal to the curve (3.4) may be disregarded; it will suffice to investigate stability along the direction of the tangent to the curve (3.4).

In the neighbourhood of the curve (3.4), we introduce a new variable x which will be measured from the normal to the curve (3.4) at the point L_0 along the direction of the tangent to the curve (3.4) at the point L_0. The directing vector of this tangent will be denoted by l_0 and

$$dx = (l_0 dL). \tag{3.5}$$

Deviation of the integral curve from the straight line on which x is measured, along the normal to this direction is be small (of order of μ or $O(\epsilon)$) in the ϵ-neighbourhood of the point L_0. Thus, using the Laplace theorem, we have to investigate stability with respect to the one variable x only.

5.4   Construction of Perturbed Lyapunov Function and Stability of Point – Like Model of Three – Body Problem

To investigate stability in the three – body problem, we will apply Theorem 4.3 which, using the perturbed Lyapunov function technique gives an estimate of the time interval where solutions in a part of the variable remain close to the point whose stability is under investigation.

In the three – body problem we fix ϵ, $0 < \epsilon$ and find $\eta(\epsilon)$, $T(\epsilon,\mu)$ and $\mu_0(\epsilon)$ such that the solutions which satisfy the condition $|L - L_0| < \eta$ in the variables L at the initial moment $t = 0$, will remain in the ϵ-neighbourhood for all t, $0 < t < T(\epsilon,\mu)$ and $\mu < \mu_0$, i.e. the inequality $|L - L_0| < \epsilon$ will be satisfied. The integrals (2.7) and (2.8) permit us to conclude that, while the variables L change within the ϵ-neighbourhood, the changes of the variables σ are of the same order.

As the non – perturbed Lyapunov function we take

$$v_0(L) =\| x \| . \tag{4.1}$$

We look for the perturbed Lyapunov function of the form

$$v = v_0(L) + \mu v_1(L, \lambda, \sigma, \omega, \epsilon). \tag{4.2}$$

Let us differentiate v and make use of Eq. (2.6):

$$\dot{v} = \dot{x}\,\mathrm{sign}x + \mu\frac{\partial v_1}{\partial \lambda}(n + \mu\sum_{k,p}\frac{\partial h_{k,p}}{\partial L}\cos\theta) +$$

$$(4.3)$$

$$+\mu\frac{\partial v_1}{\partial L} + \mu\frac{\partial v_1}{\partial \sigma}\sigma + \mu\frac{\partial v_1}{\partial \omega}\omega.$$

Let us represent the series

$$\dot{x}\text{sign}x = \mu\sum_{k,p}h_{k,p}(k\,l)\sin\theta\text{sign}x \qquad (4.4)$$

as a sum of two series

$$\dot{x}\text{sign}x = \mu\tilde{X} + \mu R_\epsilon, \qquad (4.5)$$

$$\mu R_\epsilon = \mu\overline{\sum}_{k,p}h_{k,p}(\overline{k}\,\overline{l})\sin\theta\text{sign}x. \qquad (4.6)$$

R_ϵ contains all the term whose resonance rays lie in the 2ϵ-neighbourhood of the point L_0, the sum R_ϵ is denoted by an over - bar.

$$\mu\tilde{X} = \mu\widetilde{\sum}_{k,p}h_{k,p}(\overline{k}\,\overline{l})\sin\theta\text{sign}x. \qquad (4.7)$$

The sum $\tilde{X}$ (with an over - tilda) contains oscillating terms as well as resonance terms belonging to k_0. Let us choose η ($\eta < \epsilon$) and $\sigma > 0$ satisfying the condition

$$\sigma < \frac{1}{2}(\epsilon - \eta). \qquad (4.8)$$

Let us require that the function v_1 satisfies the equation

$$\frac{\partial v_1}{\partial \lambda}n = \text{sign}x\widetilde{\sum}_{k,p}h_{k,p}(\overline{k}\,\overline{l})\sin\theta. \qquad (4.9)$$

Since

$$|\nabla\Lambda(L)| = \sqrt{k_1^2\left(\frac{3M_1}{L_1^4}\right)^2 + k_2^2\left(\frac{3M_2}{L_2^4}\right)^2} \neq 0,$$

the denominators $kn = \Lambda(L)$ which appears after integrating Eq. (4.9) are bounded from below by quantities of the order of η or ϵ, $\eta < |x| < \epsilon$. Therefore, the function v_1 is bounded for $\eta < |x| < \epsilon$ and has the order of magnitude

$$v_1 \sim O(\eta^{-1})\widetilde{\sum}_{k,p}|h_{k,p}|. \qquad (4.10)$$

By choosing μ_0 sufficiently small, it is possible to make the perturbation μv_1 of the Lyapunov function smaller than $\frac{1}{2}\sigma$ for all $\mu < \mu_0$. Then the relation (4.3) combined with Eqs. (2.6) and (4.9) give

$$\dot{v} = \mu R_\epsilon + O\left(\frac{\mu^2}{\eta}\right). \qquad (4.11)$$

For the time interval $[0, T]$ on which the solution emerging from the η-neighbourhood will remain (in the variables L), in the ϵ-neighbourhood, we have, according to Theorem 4.3, the estimate

$$T \sim \frac{\sigma}{2\mu(R_\epsilon + O(\mu^2/\eta))}. \qquad (4.12)$$

Let us estimate R_ϵ whose value for fixed ϵ is determined by the convergence rate of the series (2.3). The estimate of R_ϵ will be obtained by the technique proposed in Chapter 4 for a single – frequency system. The vectors k which correspond to the resonance rays passing through the 2ϵ-neighbourhood of the point L_0 , will be denoted by k_ϵ. To estimate the angle between the vectors k_0 and K_ϵ, we introduce the unit vector

$$\kappa = \{\frac{k_1}{|k|}, \frac{k_2}{|k|}\}$$

and consider the combinative frequency

$$\Lambda(L) = |k|(\kappa n(L)). \qquad (4.13)$$

At the point L_0.

$$\kappa_0 n(L_0) = 0. \qquad (4.14)$$

On the other resonance lines

$$\kappa_\epsilon n = (\kappa_0 + \Delta\kappa)(n_0 + \Delta n) = 0. \qquad (4.15)$$

Hence, taking (4.14) into account, we obtain an expression for the angle between the vectors κ_ϵ and κ_0 :

$$\Delta\kappa \simeq \frac{-\kappa_e \Delta n(L)}{|n_0|}. \qquad (4.16)$$

The frequencies $n(L)$ remain unchanged along the resonance rays. The variable x is measured along any straight line which is oblique relative to the resonance ray; therefore $\Delta n = O(\epsilon)$ on the boundary of the 2ϵ-neighbourhood of the point L_0.

On the other hand, when the angle β between two integer – valued vectors on the plane is small, we have the relation

$$\sin \beta = \frac{k_{10}k_2 - k_{20}k_1}{|k_0|\,|k|} \simeq \beta. \qquad (4.17)$$

The modulus of the numerator of this expression cannot be less than one for vectors different from k_0.

Comparing the angles (4.16) and (4.17), we obtain the estimate of the minimal vector $k_{\epsilon\,min}$ which lies on the boundary of the angular ϵ-neighbourhood of the vector k_0:

$$\frac{1}{|k_0|\,|k_{\epsilon\,min}|} \simeq \frac{|\kappa_0\Delta n(L)|}{|n_0|}, \qquad (4.18)$$

where k_0 is a fixed vector, $\Delta n(L) = O(\epsilon)$; therefore, when ϵ decreases, the value $|k_{\epsilon\,min}|$ increases as ϵ^{-1}. The remainder R_ϵ contains the terms of the series belonging to vectors k_ϵ, the latter being not smaller than $k_{\epsilon\,min}$. Thus the relation (4.18) enables us to estimate the remainder R_ϵ.

Since the Hamiltonian function is analytic, and since the estimate (4.18) is satisfied, we have

$$R_\epsilon \sim \exp(-a/\epsilon).$$

Taking $\sigma \sim \epsilon$, $\eta \sim \epsilon$ as well, we obtain the relation (4.12) in the form

$$T \sim \epsilon\mu^{-1}\left(exp\left(\frac{-a}{\epsilon}\right) + \frac{\mu}{\epsilon}\right)^{-1}.$$

It should be noted that, in a given three – body problem, the parameter μ has small but fixed value, and the length T of the time interval, on which the trajectory remains in the ϵ-neighbourhood of the point L_0

cannot be increased by making this parameter smaller. However this
can be done by diminishing R_ϵ.

The perturbation μF_1 of the Hamiltonian function can be expanded
([96], [95]) into the series (2.3) in power of small eccentricities and
inclinations. Therefore, the order of a term belonging to $k = \{k_1, k_2\}$
is not smaller than $||k_1| - |k_2||$. Hence, for small eccentricities and
inclinations and for sufficiently small ϵ, the value of R_ϵ in the relation
(4.11) may be a small quantity of order higher than μ. Let us now
assume that $R_\epsilon = O(\mu^2)$, and let us construct another approximation
of the function v. We look for a perturbed Lyapunov function in the
form

$$v = v_0(x) + \mu v_1(x) + \mu^2 v_2(x), \qquad (4.19)$$

where v_1 has been already defined as a solution of Eq. (4.9). Let us
differentiate v using Eq. (2.6) and taking (4.9) into account; we obtain

$$\begin{aligned}
\dot{v} = &\mu R_\epsilon + \mu^2 \frac{\partial v_1}{\partial \lambda}\frac{\partial F_1}{\partial L} - \mu^2 \frac{\partial v_1}{\partial \sigma}\frac{\partial F_1}{\partial \omega} + \\
&+ \mu^2 \frac{\partial v_1}{\partial \omega}\frac{\partial F_1}{\partial \sigma} - \mu^2 \frac{\partial v_1}{\partial L}\frac{\partial F_1}{\partial \lambda} + \mu^2 \frac{\partial v_2}{\partial L}\left(n + \mu \frac{\partial F_1}{\partial L}\right) - \\
&- \mu^3 \frac{\partial v_2}{\partial \sigma}\frac{\partial F_1}{\partial \omega} + \mu^3 \frac{\partial v_2}{\partial \omega}\frac{\partial F_1}{\partial \sigma} - \mu^3 \frac{\partial v_2}{\partial L}\frac{\partial F_1}{\partial \lambda}.
\end{aligned} \qquad (4.20)$$

Here, we multiply the series containing the derivative of the function v_1,
then we identify the resonance term using the above-mentioned method,
and define v_2 as a solution of the equation

$$\frac{\partial v_2}{\partial \lambda}n = \frac{\widetilde{\partial v_1}}{\partial L}\frac{\widetilde{\partial F_1}}{\partial \lambda} - \frac{\widetilde{\partial v_1}}{\partial \lambda}\frac{\widetilde{\partial F_1}}{\partial L} + \frac{\widetilde{\partial v_1}}{\partial \sigma}\frac{\widetilde{\partial F_1}}{\partial \omega} - \frac{\widetilde{\partial v_1}}{\partial \omega}\frac{\widetilde{\partial F_1}}{\partial \sigma}. \qquad (4.21)$$

Thus the estimate of T becomes

$$T(\epsilon, \mu) \sim \frac{\sigma(\epsilon)}{2\mu\left(R_\epsilon + O\left(\frac{\mu^2}{\eta^2}\right)\right)}. \qquad (4.22)$$

Construction of higher approximations in a given problem may become impossible, since ϵ is bounded from below by η, which must be

large enough to ensure that the η-neighbourhood of the resonance line contains the initial data of the problem.

The result just obtained can be applied to the system the Sun – Jupiter – Saturn. The parameters of the system are: the ratio of masses μ, $\mu \sim 10^{-3} - 10^{-4}$; the resonance vector $k = -2.5$; e and i are eccentricities and inclinations, $e \sim 10^{-1}$, $i \sim 10^{-1}$. We choose the quantities ϵ, σ, and η of the order of $10^{-1} - 10^{-2}$, to make the neighbouring resonances of the same order lying outside the ϵ-neighbourhood. If the vectors $\bar{k}_\epsilon$, with which R_ϵ begins, are only 2-3 times larger than the resonance vector (-2.5), then R_ϵ will be of the order $10^{-6} - 10^{-9}$. Consequently, for the time interval $T(\epsilon, \mu)$, on which the semi-axes remain in the ϵ-neighbourhood of their initial value, we have

$$T(\epsilon, \mu) \sim T_J \left(\frac{1}{\mu^2} \right) \sim 10^6 - 10^8.$$

Since the time scale here is the period of Jupiter's revolution, T is of order $10^6 - 10^8$ yrs.

It should be noted that, as when investigating the three – body problem using the generalized Lyapunov method, one can investigate stability in a problem involving more than three bodies [57].

5.5 Corrections to Force Function in Hydrodynamic Model of Planets

We will use the canonical variables introduced in Section 2.

In order to obtain corrections to the force function, we assume that the mass of a planet outside the sphere of radius r_p is proportional to the elongation parameter over the equator α. Thus, the force function must be determined for the layer of mass $M_p \alpha \bar{A}$ where M_p is the planet mass and $\bar{A}$ is the coefficient of proportionality. For the sake of simplicity, we approximate the layer mass by the sum of masses of three annular regions, one along the equator and the other two along the 45° parallel north and south of the equator.

Let us decompose this function into powers of $\bar{\mu} = r_p/R_0$. After integrating over the annular regions and summing, only the terms of the second order in μ remain within the accuracy of e^2:

$$\bar{\mu}^2 F_2 =$$
$$\frac{1}{6}\bar{\mu}^2 L_1^{-1}\alpha\bar{A}(-0.25 + 0.75\cos^2\psi\sin^2\beta)(1 + 3e\cos v), \quad (5.1)$$

where β is the inclination of the axis of rotation of the planet, $\psi = v+\pi+k_0$, and k_0 (which will be chosen later) is a constant characterizing orientation of the inclination vector of the angular momentum of the planet with respect to a static plane. In what follows it is convenient to express F_2 via the angle $\lambda = 2\pi + M$:

$$\cos^2\psi = \cos 2(\lambda - \pi - k_0) +$$
$$+2e[\cos(3\lambda - 4\pi - 2k_0) - \cos(\lambda - 2k_0)]. \quad (5.2)$$

$$\cos^2\psi\cos v =$$
$$= \frac{1}{2}[\cos(\lambda - 2\pi) + \cos(3\lambda - 4\pi - 2k_0) + \cos((\lambda - 2k_0)] + (5.3)$$
$$+e[1.5\cos(4\lambda - 6\pi - 2k_0) - \cos 2(\lambda - \pi - 2k_0) - 1 -$$
$$- \cos(2\lambda - 4\pi) - \frac{1}{2}\cos(2\pi - 2k_0)].$$

Now we write the Hamiltonian function F, taking the correction $\bar{\mu}^2 F_2$ into account

$$F = F_0 + \mu F_1 + \bar{\mu}^2 F_2, \quad (5.4)$$

where $F_0 = M_1(2L_1^2)^{-1} + M_2(2L_2^2)^{-1}$, M_1 and M_2 are constants and μF_1 is the perturbing function (2.3). It is convenient to express F_1 in terms of eccentric variables Γ and oblique variables Z:

$$F_1 = \sum A(L)\Gamma_1^{q_1} Z_1^{q_2} \Gamma_2^{q_3} Z_2^{q_4} \cos(k\lambda + p\omega) \quad (5.5)$$

Here $\sum k_i = \sum p_i$ (k and p are integer – valued vectors such that $2q_i \geq p_i$).

5.6   Theorem on Stability of Planetary Systems

As stability of a *three – body planetary system* we understand a situation when the eccentricities and inclinations of the planets remain small and the values of their semi-axes remain close to the initial values.

In [59], the following proposition was proved.

Theorem 5.1 *Let the central gravitating body in a three – body system have a much greater mass($\mu = M_p/M_S \ll 1$). Let at least one of the planets have a noticeable inclination of the rotation axis ($\beta < 1$) and elongation over the equator ($\alpha < 1$). Let also the parameter $\bar{\mu} \sim (\bar{\mu} = r_p/R_p)$.*

Then the planetary system with small eccentricities and inclinations is stable on the infinite time interval.

Proof. According to the Laplace theorem, stability along the orbital semi-axes implies that eccentricities and inclinations remain small. The Laplace theorem is based on the area integrals of the point model

$$I = L_1 + L_2 - \sigma_1 - \sigma_2 - \sigma_3 - \sigma_4 \qquad (6.1)$$

and makes it possible to investigate stability along the semi-axes L only. It is not obvious, however, whether the hydrodynamic model admits the area integrals and thus, we examine stability along the semi-axes and the variable I.

Let us write the canonical equation of motion

$$\frac{\partial L}{\partial t} = \frac{\partial F}{\partial \lambda}, \quad \frac{\partial \lambda}{\partial t} = -\frac{\partial F}{\partial L},$$

$$\frac{\partial \rho}{\partial t} = \frac{\partial F}{\partial \omega}, \quad \frac{\partial \omega}{\partial t} = -\frac{\partial F}{\partial \rho};$$

here only the phases $\lambda = \{\lambda_1, \lambda_2\}$ are fast variables.

The estimate for the small denominator $k_1 n_1(L) + k_2 n_2(L)$, with $n = \dfrac{\partial f_0}{\partial L}$, for two fast phases λ can be obtained as in Section 3, where stability of the three – body problem over a finite time interval was investigated in the framework of the point model.

Since the function $F_0 + \mu F_1 + \bar{\mu}^2 F_2$ is an integral of motion, the integral curve on the plane L remains close to the curve $F_0(L) = F_0(L_0)$.

Let us examine a point L_0, being the intersection of the curve $F_0(L) = F_0(L_0)$ and the resonance ray

$$k_1 n_1(L) + k_2 n_2(L) = 0,$$

$$(6.2)$$

$$\frac{L_2}{L_1} = \left(\frac{M_2}{M_1}\right)^{\frac{1}{3}} \left(-\frac{k_2}{k_1}\right)^{\frac{1}{3}},$$

and let us introduce the coordinate x along the tangent to the curve $F_0(L) = F_0(L_0)$ at the point L_0.

To investigate stability, we construct the perturbed Lyapunov function for the system (3.2)

$$v = v_0 + \mu v_1 + \mu^2 v_2 + \mu^3 v_3,$$

$$(6.3)$$

which satisfies the condition of Theorem 2.1 (or 4.2).

We take

$$v_0 = |x| + |I - I_0|$$

as the unperturbed Lyapunov function, since it is not obvious if the quantity I is an integral of the system (6.2).

We differentiate the function (6.4) using Eqs. (6.2) (v_3 is independent of λ), omitting terms of order higher than $\mu\bar{\mu} \sim \mu^3$. Consequently, we find v_1, v_2, v_3 and obtain the negative mean of $\dot{v}$.

From the series for the derivative $\dfrac{\partial F_1}{\partial \lambda}$ we single out all the terms whose resonance line lie in the 2ϵ-neighbourhood of the point L_0, and we relate these term to R_ϵ. The remaining terms with oscillating harmonics will be marked by an over - tilda; the terms belonging to k_0 will be related to the oscillating terms. We choose $\epsilon > 0$ and $\eta > 0$ $(\eta < \epsilon)$ and $\sigma < \frac{1}{2}(\epsilon - \eta)$. We require the function v_1 to satisfy the equation

$$\frac{\partial v_1}{\partial \lambda}\left[\frac{\partial F_0}{\partial L}\right] + \frac{\partial v_1}{\partial \gamma}\left[\mu\overline{\frac{\partial F_0}{\partial \Gamma}}\right] + \frac{\partial v_1}{\partial z}\left[\mu\overline{\frac{\partial F_0}{\partial Z}}\right] =$$

$$= \text{sign}x\frac{\partial x}{\partial L}\overline{\frac{\partial F_1}{\partial \lambda}} - \text{sign}(I - I_0)\frac{\bar{\mu}^2}{\mu}\frac{\partial \bar{F}_2}{\partial \gamma}, \qquad (6.4)$$

where $\overline{\dfrac{\partial F_0}{\partial \Gamma}}$ and $\overline{\dfrac{\partial F_1}{\partial Z}}$ describe secular perturbations in the classical theory of secular perturbations. These perturbations do not contain λ and they are quadratic with respect to eccentricities and inclinations, $\dfrac{\partial \bar{F}_2}{\partial \gamma}$ does not contain λ. After integrating along the secular perturbation trajectories we obtain

$$v_1 = \Phi_1 + \frac{\bar{\mu}^2}{\mu^2}\bar{\Phi}_1,$$

where Φ_1 contains denominators of the order ϵ, η, while $\bar{\Phi}_1$ has a factor $\frac{\bar{\mu}^2}{\mu^2}$, since the evolution rate of the secular perturbations has the order μ.

Now let v_2 be determined from the equation whose coefficients on the left-hand side (LHS) coincide with those on the LHS of Eq. (6.5), while its RHS $\phi = \phi' + \phi'' + \phi'''$ consists of the following term (it is easy to show that since the system is Hamiltonian, all ϕ terms depend on λ)

$$\phi' = \frac{\partial \Phi_1}{\partial L}\frac{\partial F_1}{\partial \lambda} - \frac{\partial \Phi_1}{\partial \lambda}\frac{\partial F_1}{\partial L} + \frac{\partial \Phi_1}{\partial \Gamma}\frac{\partial F_1}{\partial \gamma} - \frac{\partial \Phi_1}{\partial \gamma}\frac{\partial F_1}{\partial \Gamma} +$$

$$+\frac{\partial \Phi_1}{\partial Z}\frac{\partial F_1}{\partial z} - \frac{\partial \Phi_1}{\partial z}\frac{\partial F_1}{\partial Z}. \qquad (6.5)$$

$$\phi'' = 2\text{sign}(I - I_0)\left(\frac{\partial F_2}{\partial \lambda} - \frac{\partial \bar{F}_2}{\partial \gamma}\right)\frac{\bar{\mu}^2}{\mu^2}. \qquad (6.6)$$

Here, all terms of the form $\cos(2\lambda + 2\gamma + 2k_0)$ cancel and only terms for which $\sum p \neq \sum k$ remain;

$$\phi''' = \frac{\bar{\mu}^2}{\mu^2}\left[-\frac{\partial \bar{\Phi}_1}{\partial \gamma}\frac{\partial \tilde{F}_1}{\partial \Gamma} + \frac{\partial \bar{\Phi}_1}{\partial \Gamma}\frac{\partial \tilde{F}_1}{\partial \gamma}\right]. \qquad (6.7)$$

Here, the second-order terms of $\dfrac{\partial \bar{F}_1}{\partial \Gamma}$ are grouped on the LHS of Eq. (6.5), while term of order 4, i.e. e^4, i^4 and higher, are related to R_ϵ, bearing in mind that ϕ_1 has also a factor e^2. This can be done under assumption that e^4, i^4 do not exceed μ. While $|I - I_0| < \epsilon$, $e^2 L$ and $i^2 L$ as well as ΔL are quantities of the order ϵ (this assumption correlates the quantities μ and ϵ). Having singled out, as before, the resonance terms in R_ϵ, we find v_2 consisting of v_2', v_2'' and v_2''', respectively.

Let us consider the leading term in $\dot{v}$ which ensures that the mean is negative:

$$\mu \bar{\mu}^2 \overline{\dfrac{\partial v_1}{\partial \Gamma} \dfrac{\partial F_2}{\partial \gamma}} =$$

$$= -\mu \bar{\mu}^2 \mathrm{sign} x \dfrac{\partial x}{\partial L} \cdot 2 \cdot \dfrac{1}{6} \alpha A A_2 \cos^2(2\lambda + 2\gamma). \qquad (6.8)$$

According to Eq. (5.1)-(5.3), F_2 in fact contains only one term having the property (5.5). This property holds always for F_1, hence it holds for v_1. Thus, only the product (6.9) ensures negativeness with a corresponding choice of k_0 ($k_0 = \pi/4$ for $|x| > \eta > 0$). All other term of the form

$$\mu \bar{\mu}^2 \dfrac{\partial v_1}{\partial L} \dfrac{\partial F_2}{\partial \lambda}, \quad \mu \bar{\mu}^2 \dfrac{\partial v_1}{\partial \gamma} \dfrac{\partial F_2}{\partial \Gamma}$$

are e^2 times smaller than the RHS of (6.9).

We now consider

$$-\dfrac{\partial v_2'}{\partial \lambda} \dfrac{\partial F_1}{\partial L} + \dfrac{\partial v_2'}{\partial L} \dfrac{\partial F_1}{\partial \lambda} + \dfrac{\partial v_2'}{\partial \Gamma} \dfrac{\partial F_1}{\partial \gamma} -$$

$$\qquad (6.9)$$

$$-\dfrac{\partial v_2'}{\partial \gamma} \dfrac{\partial F_1}{\partial \Gamma} + \dfrac{\partial v_2'}{\partial Z} \dfrac{\partial F_1}{\partial z} + \dfrac{\partial v_2'}{\partial z} \dfrac{\partial F_1}{\partial Z}.$$

Here slow oscillations (not containing λ) can appear, but such combinations are formed from factors satisfying the conditions (5.5), thus, the results must satisfy the condition $p_1 + p_2 + p_3 + p_4 = 0$. To compensate such terms, we determine v_3 from the equation

$$\frac{\partial v_3}{\partial \gamma}\left(-\frac{\partial \bar{F}_1}{\partial \Gamma}\right) + \frac{\partial v_3}{\partial z}\left(-\frac{\partial \bar{F}_1}{\partial Z}\right) = \text{slowly oscillating terms.} \quad (6.10)$$

If $p_1 + p_3 = 0$, then $p_1 g_1 + p_2 g_2 \neq 0$ for the eccentricity terms for two planets, since g_1 and g_2 are different and real roots of the secular equation in the theory of secular perturbation. Only in the case of oblique variable does the condition $p_2 + p_4 = 0$ yield $(p_2 + p_4)\Omega_0 = 0$; the corresponding harmonics contain sin and they are reduced to zero, since the angular velocity for the motion of a node along the Laplace plane is similar for both planets. The same holds true for a mixed combination: the denominator of v_3 is different from 0 and has the order $\frac{1}{\mu}$, hence the function v_3 is of the same order.

It is obvious that the terms related to v_3 have an order higher than $\mu^3(e^2, i^2)$, hence it follows that they are smaller than the mean (6.9). It is easy to establish that the expression

$$\bar{\mu}^2 \mu \left[-\frac{\partial v_2''}{\partial \lambda}\frac{\partial F_1}{\partial L} + \frac{\partial v_2''}{\partial \Gamma}\frac{\partial F_1}{\partial \gamma} - \frac{\partial v_2''}{\partial \gamma}\frac{\partial F_1}{\partial \Gamma}\right] \quad (6.11)$$

is small as compared to the leading term in (6.9), because the slow harmonics contain as factors the quantities e and i in a power not less than 2. The same is true for expressions of the form (6.12) which contain v

According to the estimate obtained in Section 3 of Chapter 4, the smallest k determining the value of resonance terms in R_ϵ has the order $k^{-1} = O(\epsilon)$, i.e. it increases rather rapidly when ϵ decreases. Thus, all the conditions of Theorem 4.2 on stability are satisfied: the average (6.9) in the annulus $\eta < |x| < \epsilon$ is strictly negative, all non – oscillating terms in $\dot{v}$ are smaller than the mean by one order of magnitude. In accordance with the theorem, the point L_0 is stable over an infinite time interval in the variables L along the semi-axis. We have also proved conservation of I, the area integrals, hence stability relative to eccentricities and inclinations follows.

It is known ([91]) that the planets of the Solar system form groups which are resonance – related. This phenomenon is comprehensible if in one group there is a planet which is oblate and has a noticeably inclined rotation axis. Then this planet play a stabilizing role and the energy exchange is most intensive when there is a resonance.

5.7   Evolution of Planetary Orbits

1. The hydrodynamic model of planets, suggested above and used as a basis for studying the stability of a planetary system with small eccentricities and inclinations, will be used to obtain equations which describe evolution of the parameters – which are no longer assumed to be small (such as eccentricities, inclinations, etc.) – of a planetary system under the influence of the oblate form of both the planets and the central body, and also of the inclination of their rotational axes. Therefore, we study evolution of the Kepler orbit, its size, eccentricity and inclination, with evolution of the parameter depending on the ratio of the attracting bodies to the distance between them ([56]).

2. It is convenient to use the Cartesian co-ordinates to describe the perturbing function R, which accounts for the asymmetry of the attracting bodies. The system will be selected as follows: the (x, y) plane coincides with the static plane of the planetary system; the unit vector directed along the rotational axis of one planet has coordinates $(O, \sin\beta, \cos\beta)$; the coordinates of the central body are $(0, 0, 0)$, while those of the planet are $(x, y, z,)$. Both gravitating bodies will be assumed to be oblate ellipsoids with the eccentricities e_c and e_p, where $e_c = (a_c^2 - r_c^2)/a_c^2$, $e_p = (a_p^2 - r_p^2)/a_p^2$, a_p and a_c are the equatorial radii, r_c and r_p are the polar radii, the subscript c stands for the central body, and p for the planet. It is convenient to expand into the Legendre polynomial ([107]) retaining only terms of the order $(a_c/r)^2$. In what follows, we use the conventional notation for the orbital elements

([107]): a, e, i, ω, π, Ω, ϵ, v, v, $p = a(1 - e^2)$, $u = v + \omega$, $\omega = \pi - \Omega$. We introduce the notation

$$C_c = \frac{e_c^2}{5 \cdot 2}\left(\frac{a_c}{p_0}\right)^2, \quad C_p = \frac{e_p^2}{5 \cdot 2}\left(\frac{a_p}{p_0}\right)^2, \quad \bar{M} = \gamma M_c M_p,$$

and we obtain

$$R = \bar{M}p_0^2\{\frac{3}{r^5}[C_c z^2 + C_p(x^2 + y^2 \cos^2\beta - yz \sin 2\beta)] - \frac{C_c + c_p}{r^5}\} \quad (7.1)$$

with p_0 being the initial value of p. Here, terms of the order $\sin^4\beta$ are omitted, the assumption being that

$$C_c > C_p \sin^4\beta.$$

3. Here we present the equations for oscillating terms of the elliptic orbit in the Lagrange form [107], where, as an independent variable, we took the genuine anomaly:

$$C\frac{da}{dv} = +\frac{2}{na^2}\frac{\partial R}{\partial \epsilon}r^2 \quad (7.2)$$

$$C\frac{de}{dv} = -\frac{\sqrt{1 - e^2}}{na^2 e}\frac{\partial R}{\partial \pi}r^2 - \frac{e\sqrt{1 - e^2}}{1 + \sqrt{1 - e^2}}\frac{1}{na^2}\frac{\partial R}{\partial \epsilon}r^2, \quad (7.3)$$

$$C\frac{di}{dv} = -\frac{\operatorname{cosec} i}{na^2\sqrt{1 - e^2}}\frac{\partial R}{\partial \Omega}r^2 - \frac{\tan\frac{1}{2}i}{na^2\sqrt{1 - e^2}}\left(\frac{\partial R}{\partial \pi} + \frac{\partial R}{\partial \epsilon}\right)r^2 \quad (7.4)$$

$$C\frac{d\Omega}{dv} = \frac{\operatorname{cosec} i}{na^2\sqrt{1 - e^2}}\frac{\partial R}{\partial i}r^2, \quad (7.5)$$

$$C\frac{d\pi}{dv} = \frac{\tan\frac{1}{2}i}{na^2\sqrt{1 - e^2}}\frac{\partial R}{\partial i}r^2 + \frac{\sqrt{1 - e^2}}{na^2 e}\frac{\partial R}{\partial e}r^2, \quad (7.6)$$

$$C\frac{d\epsilon}{dv} = -\frac{2}{na^2}\frac{\partial R}{\partial a}r^2 + \frac{\tan\frac{1}{2}i}{na^2\sqrt{1 - e^2}}\frac{\partial R}{\partial i}r^2 +$$

$$+\frac{e\sqrt{1 - e^2}}{1 + \sqrt{1 - e^2}}\frac{1}{na}\frac{\partial R}{\partial e}r^2. \quad (7.7)$$

Here $C = r^2\dot{v}$ is the integral of areas. All the variables a, e, i, Ω, π, ϵ are slow, the small parameter is the squared ratio of the size of the bodies to the distance between them.

Now we describe a rather cumbersome procedure for averaging Lagrange equation over the anomaly v, using Eq. (7.2) as an example. To obtain the derivative $\dfrac{\partial R}{\partial \epsilon}$, one must first differentiate R with respect to x, y, z and then the variables

$$x = r(\cos u \cos \Omega - \sin u \sin \Omega \cos i)$$
$$y = r(\cos u \sin \Omega + \sin u \cos \Omega \cos i) \qquad (7.8)$$
$$z = r \sin u \sin i$$

must be differentiated with respect to r and u using the derivatives [107]

$$\frac{\partial r}{\partial \epsilon} = \frac{ae \sin v}{\sqrt{1 - e^2}}, \qquad \frac{\partial u}{\partial \epsilon} = \frac{a^2 \sqrt{1 - e^2}}{r^2} \qquad (7.9)$$

and the expression

$$r = \frac{p}{1 + e \cos u}.$$

Thus, on the RHS of Eq. (7.1), which is periodic in v, the constant component of the Fourier series must be separated out.

Let us write the mean of $r^2 \dfrac{\partial R}{\partial \epsilon}$ with the accuracy $C_p \sin^2 i$:

$$r^2 \overline{\frac{\partial R}{\partial \epsilon}} = \bar{M} p_0^2 \left\{ \frac{3(-5)}{4p^2} \frac{ae}{\sqrt{1 - e^2}} \left[\frac{3}{5} C_c e \sin 2\omega \sin^2 i - \right. \right.$$
$$- C_p e \sin^2 \beta \sin 2\pi - C_p e \sin i \sin 2\beta \sin(2\omega + \Omega) \Big] +$$
$$+ \frac{ae}{\sqrt{1 - e^2}} \frac{3e C_p}{2p^2} \left[- \sin^2 \beta \sin 2\pi - \sin 2\beta \sin i \sin \Omega \cos 2\omega \right] +$$
$$+ \frac{9e^2 a^2}{4p^3} C_p \left[- \sin^2 \beta \sin 2\pi + \sin i \sin 2\beta \sin(\Omega + 2\omega) \right] + \qquad (7.10)$$
$$\left. + \frac{9e^2 a^2}{4p^3} \sqrt{1 - e^2} \sin^2 i \sin 2\omega \right\}.$$

Let us also write the mean over the angle v of $r^2 \dfrac{\partial R}{\partial e}$

$$r^2 \overline{\frac{\partial R}{\partial e}} = \frac{\bar{M} p_0^2 ae}{p^2} \left\{ -3(C_c + C_p) + \frac{3}{4} C_c \sin^2 i \cos 2\omega + \right.$$

$$+C_p \sin^2 \beta \left[\frac{15}{4} \cos 2\Omega \cos 2\omega - \frac{9}{2} \cos 2(\Omega - \omega) \right] + \qquad (7.11)$$

$$+C_p \sin i \sin 2\beta \left[\frac{15}{4} \cos 2\omega \cos \Omega - \frac{9}{2} \cos(2\omega - \Omega) \right] \Big\} .$$

4. Let us average over v, Eqs. (7.4) and (7.5). First we note that in Eq. (7.4) there is a term $r^2(\frac{\partial R}{\partial \pi} + \frac{\partial R}{\partial \epsilon})$, whose mean is equal to zero; this is easy to see:

$$\frac{\partial r}{\partial \pi} = -\frac{ae \sin v}{\sqrt{1 - e^2}}, \quad \frac{\partial u}{\partial \pi} = -\frac{a^2 \sqrt{1 - e^2}}{r^2} + 1,$$

$$C\frac{\overline{d\Omega}}{dv} = \frac{\bar{M}p_0^2}{p} \sin 2i \left\{ \frac{1}{2}(C_c - C_p) + C_p \cos \Omega[(\sin^2 \beta \cos \Omega - \right.$$

$$\left. - \frac{1}{4}\sin 2\beta) + \frac{1}{2}\sin 2\beta \tan i] \right\} \frac{\mathrm{cosec}\, i}{na^2\sqrt{1 - e^2}}. \qquad (7.12)$$

$$C\frac{\overline{di}}{dv} = \frac{\bar{M}p_0^2}{p} 3C_p \left\{ \frac{1}{4} \sin \Omega \sin 2\beta \sin 2i - \right.$$

$$\left. - \sin^2 \beta \sin^2 \frac{i}{2} \sin 2\Omega(1 + \cos i) \right\} \frac{-\mathrm{cosec}\, i}{na^2\sqrt{1 - e^2}}. \qquad (7.13)$$

According to the averaging principle [12], the evolution of the orbital parameters under influence of the above – mentioned asymmetry in the mass distribution of the interacting bodies over an asymptotically large time interval of order $\frac{1}{C_p}$ or $\frac{1}{C_c}$, is determined by Eq (7.2)-(7.7), their RHS being averaged over v . It is possible, however, to single out some interesting situations in which the tendency of orbital parameter to change may be identified qualitatively.

It is essential that the signs of the RHS of the averaged Eqs. (7.12) and (7.13) depend only on the variables Ω (the longitude angle) and i (inclination); this fact makes it possible to trace the evolution of these variables on the phase plane (i, Ω). The RHS of Eq. (7.12) is a quadratic trinomial in $\cos \Omega$ with the roots

$$\cos \Omega =$$

$$\frac{1}{\sin^2 \beta} \left[\frac{1}{4} \sin 2\beta (\tan i - \frac{1}{2}) \pm \sqrt{\sin^2 2\beta (\tan i - \frac{1}{2}) - \gamma_0 \sin^2 \beta} \right],$$

$$\gamma_0 = \frac{1}{2} \left(\frac{C_c}{C_p} - 1 \right). \tag{7.14}$$

It is clear that there exists a region of variation of the parameters β, γ_0 and the variable i, where there are roots not exceeding 1 in modulus. One of the roots is on the increasing branch of $\cos \Omega$, while the other is on the decreasing branch; one of these roots is stable, while the other is not. Let us examine the situation in which the stable root is in the segment $[0, \frac{1}{2}\pi]$ and the inclination i is small ($\sin 2i > \sin^2 \frac{i}{2}$). Then on the RHS of Eq. (7.12) the first term, which contains $\sin \Omega > 0$ is dominant; this means that the sign of the RHS of Eq. (7.12) is minus and that i decreases. For larger i, the second term which contains $\sin 2\Omega$ is dominant; then i increases while such a regime continues. This explains why the inclinations of planets are small in a real system.

5. Now, let us write the averaged Eqs. (7.3) and (7.6) for the eccentricity e of the angle π. We must take into account that the first term in (7.3) contains e^{-1}, while the second contains e . Furthermore, as has been mentioned before, $r^2 \dfrac{\overline{\partial R}}{\partial \pi} = -r^2 \dfrac{\partial R}{\partial \epsilon}$. Therefore, we retain only the first leading term when e is small, i.e. the expression (7.10) with the minus sign:

$$C\frac{\overline{da}}{dv} = \frac{2}{na^2} \frac{\overline{\partial R}}{\partial \epsilon} r^2$$

$$\tag{7.15}$$

$$C\frac{\overline{de}}{dv} = \frac{\sqrt{1 - e^2}}{na^2 e} \frac{\overline{\partial R}}{\partial \pi} r^2$$

Let us also write the averaged equation for the angle π

$$C\frac{\overline{d\pi}}{dv} = \frac{\sqrt{1 - e^2}}{na^2 e} \frac{\overline{\partial R}}{\partial e} r^2 \tag{7.16}$$

From Eq. (7.10) we retain only the terms of the first order with respect to the eccentricity e , assuming that $e < 1$. Then Eqs. (7.15) and (7.16) take the following form:

$$C\frac{\overline{d\pi}}{dv} = \bar{M}\frac{p_0^2}{p^2 na}\left\{-3(C_c + C_p) + C_c\frac{3}{4}\sin^2 i \cos 2\omega + \right.$$

$$+ C_p \sin^2\beta\left[\frac{15}{4}\cos 2\Omega\cos 2\omega - \frac{9}{2}\cos 2(\Omega - \omega)\right] + \tag{7.17}$$

$$\left. + C_p \sin i \sin 2\beta\left[\frac{15}{4}\cos 2\omega\cos\Omega + \frac{9}{2}\cos(2\omega - \Omega)\right]\right\},$$

$$C\frac{\overline{de}}{dv} = \frac{\bar{M}p_0^2 e}{p^2 na}C_p\left\{-\frac{9}{2}\sin^2\beta\sin 2(\omega + \Omega) + \right.$$

$$\left. + \frac{3}{2}\sin 2\beta\sin i\sin 2\omega\cos\Omega\right\}. \tag{7.18}$$

For the semi – axis a, Eq. (7.15) is similar to that for e. Eqs. (7.3) and (7.6) obtained by averaging (7.16) and (7.17) can be also analyzed on the phase plane (e, π) or (e, ω), provided that the parameters of the system admit an asymptotically stable regime $\Omega = \Omega_0$ described in Section 6. Let us divide Eq. (7.18) by (7.17) and find the derivative $\frac{de}{d\pi}$ or (if $\Omega = \Omega_0$) $\frac{de}{d\omega}$. We assume that the term $3(C_c + C_p)$ on the RHS of Eq. (7.18) is larger than the other terms containing $\sin^2 i$, $\sin^2\beta$ etc. We expand the RHS in powers of $C_p\frac{\sin^2\beta}{3(C_c+C_p)}$, $C_c\frac{\sin^2 i}{3(C_c+C_p)}$ and so on, retaining the first order terms only. As a result, we get

$$\frac{de}{d\omega} = -e\frac{9}{16}\frac{C_p^2}{(C_c + C_p)^2}\left\{-\frac{C_c}{C_p}\sin^2 i\sin^2\beta\sin 2\Omega_0 - \right.$$

$$- \frac{7}{2}\sin^4\beta\sin 4\Omega_0 + \sin^2\beta\sin 2\beta\sin i\left(-\frac{5}{2}\sin\Omega_0 - \cos\Omega_0 \tag{7.19}\right.$$

$$\left. - \cos 3\Omega_0 + \frac{7}{2}\sin 3\Omega_0\right) + \sin^2 i\sin^2 2\beta\sin 2\Phi_0\right\} +$$

$$+ A\cos 2\omega + B\sin 2\omega.$$

The direction of evolution of the eccentricity e (and, similarly, of the semi-axis a) is determined by the sign of the sum containing Ω_0 in Eq. (7.19), while the terms $A\cos 2\omega$ and $B\sin 2\omega$ yield 0 after integrating over the period and determine the oscillations of angle frequency ω.

At this point, we should note that the evolution of e and a is determined by the oblateness of both the planet and the central body, though it is possible at $C_c = 0$. Here, the situations when e decreases can be also singled out. To this end, we assume the inclination i to be small, then the second term in Eq. (7.19) becomes dominant. The parameters of the system in (7.14) may be such that, if $\sin 4\Phi_0 < 0$, then e decreases.

7. Thus, we have constructed a mathematical model describing the evolution of the orbital parameters of a planet under the influence of the oblateness of both the planet and the central body, under condition that the rotational axis of the planet remains unchanged. We have indicated the specific features of the model which explain why both inclinations and eccentricities are small in a real system. It is important that the factors relevant in our model continue to be important when a third body is present, i.e. in the three – body problem.

CHAPTER 6

Stability of Systems with Admissible Region of Motion. Stability of Gyroscope with No – Contact Suspension

In this chapter, we present a generalization of the Lyapunov second method which can be applied to two-dimensional systems with respect to slow variables, and which takes into account the known region of motions. Stability conditions containing only observed parameters of a system are established for a gyroscope with a no-contact suspension. Previously obtained results were based on a reduction of a two – frequency system to a single – frequency one by considering a narrow category of motions in the vicinity o the equilibrium position and contained unobservable gyroscope phases.

167

6.1   Estimation of Region of Motion of the System

Let us consider a system of ordinary differential equations of the form

$$\dot{z} = f(z,t) + \mu R(z,t,\mu),\qquad (1.1)$$

with

$$0 \leq \mu \leq \bar{\mu}, \quad \bar{\mu} > 0, \quad z = (x,y) \in \mathbf{R}^m \times \mathbf{R}^n$$

$\dim x = m$, $\dim y = n$.

Stability of this system is investigated under assumption that the region of the motions of the system is known. This makes it possible to relax the restrictions on frequencies necessary in order to estimate resonance frequencies.

The system of equations describing motion of a gyroscope with a no-contact suspension satisfies this condition ([80], [81]) . Therefore the approach which takes into account the fact that the region of motions is known makes it possible to establish stability conditions in terms of observable parameters only ([47], [29], [49], [48]).

The functions f and R are defined in the region

$$z \in A \times D \subset \mathbf{R}^m \times \mathbf{R}^n, \quad t \geq 0, \quad 0 \leq \mu \leq \bar{\mu},\qquad (1.2)$$

with assumption that the region A contain the origin of the space $\mathbf{R}^m$. Let us assume that the system (1.1) has in the region (1.2) the unique continuous solution $z(t, z_0, t_0, \mu)$ with the initial condition $z = z_0$ at $t = t_0$. We also assume that this solution is either defined for all $t \geq t_0$, or can be continued in the variables x to the boundary of the region A.

Let the generating system

$$\dot{z} = f(z,t)\qquad (1.3)$$

have an equilibrium position in the variables x at the point $x = 0$ and let this system have a known Lyapunov function $v(z, t)$ which is positive – definite and continuous in x at the point $x = 0$ uniformly with respect to $y \in D$, $t \geq 0$ and such that $v(0, y, t) \equiv 0$. It is also assumed that $v(z, t)$ is differentiable in the region (1.2) and

$$\frac{\partial v}{\partial t} + \frac{\partial v}{\partial z} f \leq 0.$$

Later we will relax the requirement for the derivative of the generalized Lyapunov function to be negative – definite; this will be done by taking into account additional information that the motion is restricted to a specific region only.

This property of motion will be taken into account when estimating small denominators in resonance problem.

The additional properties of the system (1.1), which can be used to analyze stability, will be expressed in terms of a real function $u = u(z, t)$ defined in the region

$$z \in A \times D, \quad t \geq 0. \tag{1.4}$$

Concerning general properties of the function u, we assume that in the region (1.4) the function u has constant sign (e.g. $u \geq 0$), at the point $x = 0$ the function is continuous in x uniformly with respect to $y \in D$, $t \geq 0$, and that in the vicinity of this point the function is differentiable with respect to z and t, with $u(0, y, t) \equiv 0$.

For fixed $\epsilon > 0$, we denote

$$Q(\epsilon) = \{\| x \| < \epsilon\}$$
$$\eta_0(\epsilon) = \sup_{z,t} u(z, t), \quad z \in \partial Q(\epsilon) \times D, \quad t \geq 0. \tag{1.5}$$

Let us introduce the sets

$$H(\eta, \epsilon) = \{x \in Q(\epsilon) : u(z, t) < \eta\}$$
$$G(\eta, \epsilon) = \{x \in Q(\epsilon) : u(z, t) = \eta\} \tag{1.6}$$

Note that for $0 < \eta < \eta_0(\epsilon)$ and $G(\eta, \epsilon) \neq \emptyset$ the set $H(\eta, \epsilon)$ contains the ball $Q(\delta)$, $\delta > 0$.

Definition 6.1 *Any set containing the set $H(\eta, \epsilon) \times D$ for some $0 <$
$\eta < \eta_0(\epsilon)$ will be called the region of motions of the system (1.1), if
there exists a number $\hat{\mu} > 0$ such that the solution $z(t, z_0, 0, \mu)$ with the
initial data $z_0 \in H(\eta, \epsilon) \times D$ and the parameter μ, $0 \leq \mu \leq \hat{\mu}(\epsilon)$ can
leave the set A_ϵ in the variables x if and only if the solution leaves in
the variable x the ball $Q(\epsilon)$.*

In particular, the region $Q(\epsilon) \times D$ is the region of motions of the
system (1.1).

Let us now formulate the lemma establishing conditions sufficient
for existence of a region of motions of the form

$$H(\eta'', \epsilon) \times D, \quad 0 < \eta'' < \eta_0(\epsilon) \tag{1.7}$$

for fixed region of initial data

$$H(\eta', \epsilon) \times D, \quad 0 < \eta' < \eta''. \tag{1.8}$$

The function $F(z, t, \mu)$, defined in the region $z \in A \times D$, $t \geq 0$,
$0 \leq \mu \leq \bar{\mu}$ and taking its values in a normed m-dimensional space is said
to belong to the class $K_z^m(\mathcal{L}_0, \sigma)$ if there exists an integrable function
$\mathcal{L}(t)$, a constant $\mathcal{L}_0$ and a non-decreasing function $\sigma(\alpha)$, $\lim_{\alpha \to 0} \sigma(\alpha) = 0$
such that

$$\| F(z'', t, \mu) - F(z', t, \mu) \| \leq \mathcal{L}(t)(\sigma(\| z'' - z' \|) + \sigma(\mu))$$

and for any t_1, t_2, $t_2 \geq t_1 \geq 0$

$$\int_{t_1}^{t_2} \mathcal{L}(t) \, dt \leq \mathcal{L}_0(t_2 - t_1).$$

If for the function $F(z, t, \mu)$ the estimate $\| F(z, t, \mu) \| \leq N$ is satis-
fied, where the integrable function $N(t)$ satisfies conditions similar to
that for the function $\mathcal{L}(t)$ (but with the constant N_0), then the function
F is said to belong to the class $M^m(N_0)$.

Lemma 6.1 *Let the right hand side of the system* (1.1) *be such that* $f \in K_z^{m+n}(\mathcal{L}_0, \sigma)$, *where* $\sigma(\alpha) \equiv \alpha$, $R \in M^{m+n}(N_0)$.

Then there exists a function $\beta(t, \mu)$ *non-decreasing in t and satisfying* $\lim\limits_{\mu \to 0} \beta(t, \mu) = 0$ *(for any fixed t) such that*

$$\| z(t, z_0, t_0, \mu) - z(t, z_0, t_0, 0) \| \leq \beta(t - t_0, \mu)$$

for any $z_0 \in A \times D$, $t \geq 0$.

Lemma 6.2 *Let us assume that conditions of Lemma 6.1 are fulfilled. The function* $u(t, z)$ *satisfies in the region* (1.4) *the inequality*

$$\frac{\partial u}{\partial t} + \frac{\partial u}{\partial z} f \leq 0.$$

while $\Phi = \dfrac{\partial u}{\partial z} R \in K_z^1(\mathcal{L}_0, \sigma)$.

Then there exists a function $\gamma(t, \mu)$, *non-decreasing in t,* $\lim\limits_{\mu \to 0} \gamma(t, \mu) = 0$ *which for any t satisfies the inequality*

$$u(t, z(t, z_0, t_0, \mu)) - u(t, z_0) \leq$$
$$\leq \mu\gamma(t - t_0, \mu) + \mu \int_{t_0}^t \Phi(z(\tau, z_0, t_0, 0), \tau, 0) \, d\tau.$$

Proof.

$$u(z, t) - u(z_0, t_0) =$$
$$= \mu \int_{t_0}^t \Phi(z(\tau, z_0, t_0, 0), \tau, 0) \, d\tau + \mu \int_{t_0}^t [\Phi(z, \tau, \mu) - \Phi(\bar{z}, \tau, 0)] \, d\tau,$$
$$\bar{z} = z(\tau, z_0, t_0, 0)$$

Combining the condition $\Phi \in K_z^1(\mathcal{L}_0, \sigma)$ and Lemma 6.1, we obtain the inequality

$$u(z, t) - u(z_0, t_0) \leq \mu \int_{t_0}^t \Phi(\bar{z}, \tau, 0) \, d\tau +$$
$$+ \mu(t - t_0)[\sigma(\beta(t - t_0, \mu)) + \sigma(\mu)].$$

Since we can assume that $\gamma(t, \mu)$ is equal to $\mu t[\sigma(\beta(t, \mu)) + \sigma(\mu)]$, Lemma 6.2 is valid.

Now we turn to study the regions of motion for the system (1.1)

Lemma 6.3 *Let there exists a non-negative function $u = u(z,t)$ such that the inequality*

$$\frac{\partial u}{\partial t} + \frac{\partial u}{\partial z} f \leq 0$$

is valid in the region (1.4). *Let the function f belong to $K_z^{m+n}(\mathcal{L}_0, \| \cdot \|)$,*

$$R \in M^{m+n}(N_0), \quad \Phi = \frac{\partial u}{\partial z} R \in K_z^1(\mathcal{L}_0, \sigma) \cap M^1(N_0).$$

Let for any $\epsilon > 0$, $\epsilon < \epsilon_0$, there exist numbers $T > 0$, $\eta_1' > 0$ such that

$$\sup_{z_0, t_0} \frac{1}{T} \int_{t_0}^{t_0+T} \Phi(\bar{z}, t, 0)\, dt < 0, \quad z_0 \in G(\eta', \epsilon) \times D, \quad t_0 \geq 0.$$

Then for any $\epsilon > 0$ there exists a region of motion of the system (1.1) *of the form* (1.7) *with the fixed region of initial data* (1.8).

Remark. If the system (1.1) is $(\hat{y}, \mu)$ - stable at the point $\hat{y} = 0$ (dim $\hat{y} \leq n$), then the integral of the function Φ can be assumed to be negative under additional condition $\| \hat{y} \| \leq \hat{e}(\epsilon)$.

Proof. We prove the lemma in the general case, taking the above remark into account. For fixed $\epsilon > 0$, we choose $\eta' > 0$, $\hat{\epsilon} > 0$, and $T > 0$ and a number $\kappa > 0$, such that

$$\frac{1}{T} \int_{t_0}^{t_0+T} \Phi(\bar{z}, t, 0)\, dt < -\kappa T \tag{1.9}$$

for any $z_0 \in G(\eta', \epsilon) \times D$, $\| y_0 \| \leq \hat{\epsilon}$, $t_0 \geq 0$. Let us fix $\eta'' > 0$ from the interval $\eta' < \eta'' < \eta_0(\epsilon)$, where $\eta_0(\epsilon)$ is defined by Eq. (1.5).

Using Lemma 6.2, we obtain the estimate of the increment of the function $u(z,t)$ along the solution of the system (1.1):

$$u(z, t) - u(z_0, t_0) \leq$$
$$\mu \int_{t_0}^{t} \Phi(\bar{z}, \tau, 0)\, d\tau + \mu \gamma(t - t_0, \mu). \tag{1.10}$$

Let $\mu_1 = (\eta'' - \eta')/T N_0$, and let a number $\mu_2 > 0$ be such that for any μ, $0 \leq \mu \leq \mu_2$, we have

$$\gamma(T, \mu) \leq \kappa T.$$

Let us consider the solution $z(t, z_0, 0, \mu)$, $z_0 \in H(\eta', \epsilon) \times D$, $\hat{y} \leq \hat{\delta}$, where a number $\hat{\delta} > 0$ is chosen such that for $t \geq 0$ the condition $\| \hat{y}(t, z_0, 0, \mu) \| \leq \hat{\epsilon}$ holds for any μ, $0 \leq \mu \leq \bar{\mu}$, which is possible since the system (1.1) is $(\hat{y}, \mu)$-stable.

Let us fix μ from the interval

$$0 < \mu < \min\{\mu_1, \mu_2, \hat{\mu}\} \tag{1.11}$$

We will show that the region $H(\eta'', \epsilon) \times D$ is the region of motions of the system.

Let us assume the contrary. Then there exist the numbers

$$t' = \sup\{t < t'' : u(z(t, z_0, 0, \mu), t) = \eta'\}$$
$$t'' = \inf\{t > 0 : u(z(t, z_0, 0, \mu), t) = \eta''\}$$

such that on the segment $t' \leq t \leq t''$ the solution in the variables x lies entirely in the ball $Q(\epsilon)$, in particular,

$$z(t', \mu) \in G(\eta', \epsilon) \times D,$$
$$z(t'', \mu) \in G(\eta'', \epsilon) \times D, \tag{1.12}$$
$$u(z(t'', \mu), t'') - u(z(t', \mu), t') = \eta'' - \eta'.$$

Two cases are possible: either $t'' - t' \leq T$ or $t'' - t' > T$. If $t'' - t' \leq T$, then from the inequality $\mu < \mu_1$ it follows immediately that

$$u(z(t'', \mu), t'') - u(z(t', \mu), t') < \eta'' - \eta',$$

which contradicts (1.12). In the second case, combining (1.9) and (1.10), we obtain the following estimate for the variation of the function u on the segment $t' \leq t \leq t' + +T$:

$$u(z(t' + T, \mu), t' + T) - u(z(t', \mu), t') \leq \mu(-kT + \gamma(T, \mu)).$$

But, according to (1.11), $\mu < \mu_2$ and, therefore, the RHS of the latter inequality is negative. This means that on the interval $t' \leq t \leq t''$ there exists a point $t' + T$ such that $z(t' + T, \mu) \in H(\eta', \epsilon) \times D$.

This result contradicts the definition of the number t'. The lemma is proved.

6.2   Stability of Systems with Known Region of Motions

We introduce the following notation for an equipotential level of the function $v(z, t)$ inside the ball $Q(\epsilon)$:

$$S(c, \epsilon) = \{x \in Q(\epsilon) : v(z, t) = c\}.$$

Theorem 6.1 *Suppose that:*

a) There exists a Lyapunov function of the system (1.3) $v = v(z, t)$ *which is positive–definite in the variables* x.

b) There exists a non-negative function $u = u(z, t)$ *such that in the domain of its definition* (1.4)

$$\frac{\partial u}{\partial t} + \frac{\partial u}{\partial z} f \leq 0.$$

c) $f \in K_z^{m+n}(\mathcal{L}_0, \| \cdot \|)$, $R \in M^{m+n}(N_0)$, $\phi, \Phi \in\in K_{z,\mu}^1(\mathcal{L}_0, \sigma) \cap M^1(N_0)$.

d) For any $\epsilon > 0$, $(\epsilon < \epsilon_0)$ *there exist* $\eta' > 0$, $\eta'' > 0$ $((\eta' < \eta'' < \eta_0(\epsilon))$, $c' > 0$ $(c' < c_0(\epsilon))$, $T > 0$ *such that*

$$\sup_{z_0, t_0} \frac{1}{T} \int_{t_0}^{t_0+T} \Phi(z(t, z_0, t_0, 0), t, 0) \, dt < 0.$$

Then the system (1.1) *is* (x, μ)-*stable.*

Remark. If the system is $(\hat{y}, \mu)$-stable at the point $\hat{y} = 0$, then condition (d) of Theorem 6.1 can be checked for $\| \hat{y}_0 \| < \hat{\epsilon}(\epsilon)$.

Proof. Let us prove the theorem in the general case, taking the above Remark into account. It follows from the conditions of the theorem that all the conditions of Lemma 6.3 are fulfilled. Therefore, if the parameter μ is chosen in accordance with the inequality (1.11), it is possible to indicate the region of motions $H(\eta'', \epsilon) \times D$ for the system (1.1), the initial data being chosen from the region $z_0 \in H(\eta', \epsilon) \times D$, $\| \hat{y}_0 \| \leq \hat{\delta}$ at $t = 0$ ($\hat{\delta} > 0$ is determined by the given $\hat{\epsilon} > 0$), η' and

η'' are fixed in accordance with condition (d) of the theorem. Let us impose on μ yet another restriction

$$\mu < \frac{c_0 - c'}{T N_0}. \tag{2.1}$$

Without loss of generality, one can assume that if $\bar{z} = z(t, z_0, t_0, 0)$

$$z_0 \in [H(\eta'', \epsilon) \cap S(c', \epsilon)] \times D, \quad \| \hat{y}_0 \| \leq \hat{\epsilon}, \quad t_0 > 0,$$

then the inequality

$$\int_{t_0}^{t_0+T} \phi(\bar{z}, t, 0)\, dt < -\kappa T, \tag{2.2}$$

which follows from the conditions of the theorem, holds for the same $\kappa > 0$ as it did in the inequality (1.9).

Let us denote

$$P(c', \epsilon) = \{x \in Q(\epsilon) : v(z, t) < c'\}.$$

Let us consider the solution $z(t, z_0, 0, \mu)$ of the system (1.1) with the initial data

$$z_0 \in [H(\eta', \epsilon) \cap P(c', \epsilon)] \times D, \quad \| \hat{y}_0 \| \leq \hat{\epsilon}.$$

Then, firstly, $\| \hat{y}(t, z_0, 0, \mu) \| \leq \hat{\epsilon}$ for $t \geq 0$ and, secondly, it follows from the properties of the region of motions that either for all $t \geq 0$ we have

$$x(t, \mu) \in H(\eta'', \epsilon), \tag{2.3}$$

or there exists a moment of time $t^* > 0$ such that for $0 \leq t < t^*$, the inclusion (2.3) is valid, while at $t = t^*$ the inclusion is $x(t^*, \mu) \in \partial Q(\epsilon)$.

To complete the proof of the theorem it is enough to show that the number t^* does not exist.

Suppose that $t^* > 0$ does exist. We define the numbers

$$t' = \sup\{t < t'' : v(z(t, \mu), t) = c'\},$$
$$t'' = \inf\{t \leq t^* : v(z(t, \mu), t) = c^0(\epsilon)\}, \tag{2.4}$$

It follows from Eq. (2.4) that

$$v(z(t'',\mu),t'') - v(z(t',\mu),t') = c^0 - c'. \tag{2.5}$$

If $t'' - t' \leq T$, then, according to (2.1), we obtain

$$v(z(t'',\mu),t'') - v(z(t',\mu),t') < c^0 - c',$$

which contradicts (2.5). If, on the contrary, $t'' - t' > T$, then, according to Lemma 6.2 and the inequality (2.2)

$$v(z(t'+T),t'+T) - v(z(t'),t') \leq \mu(-\kappa T + \gamma(T,\mu)).$$

However, the parameter μ is chosen to ensure the inequality $\gamma(T,\mu) < \kappa T$, thus, for $t = t' + T$,

$$v(z(t'+T,z_0,0,\mu),t'+T) < c',$$

which contradicts definition of the number t' given in (2.4). The theorem is proved.

6.3   Stability of Multi – Frequency Systems with Known Region of Motions

A multi – frequency system

$$\dot{z} = \mu Z(z,q), \quad \dot{q} = \omega(z) + \mu\Omega(z,q), \tag{3.1}$$

with $0 \leq \mu \leq \bar{\mu}$, $\bar{\mu} > 0$, $z = (x,y)$, $\dim x = m$, $\dim y = n$, $\dim q = r$ represents a particular case of the system (1.1). The functions Z, ω, Ω are defined in the domain

$$z \in A \times D \in \mathbf{R}^m \times \mathbf{R}^n, \quad q \in \mathbf{R}^r. \tag{3.2}$$

Using Theorem 6.1, we will formulate sufficient conditions for stability which take a special form of the system (3.1) into account. Here, we consider a case which is important for application, viz. when the function Z is a trigonometric polynomial

$$Z = \sum_{k \in K} Z_k(z) \exp(\imath kq), \quad \imath^2 = -1.$$

Here K denotes a finite number of integer - valued vectors

$$k = (k_1, \ldots, k_r), \quad kq = k_1 q_1 + k_2 q_2 + \cdots + k_r q_r.$$

Together with the system (3.1) we examine the system

$$\dot{z} = \mu Z_0(z), \tag{3.3}$$

where $Z_0(z)$ is the zero harmonic of the function Z.

Let the system (3.3) have an equilibrium position in x at the point $x = 0$. Let us assume that a Lyapunov function $v(z)$, positive – definite in the variables x at the point $x = 0$, is known and that the function

$$\phi_0 = \frac{\partial v}{\partial z} Z_0(z) \tag{3.4}$$

is negative-definite in the variable x at the point $x = 0$, i.e.

$$\sup \phi_0(z) < 0, \quad z \in \partial Q(\epsilon) \times D$$

for all $0 < \epsilon < \epsilon_0$.

If there exists an integer - valued vector $k \in K$ such that $k\omega(z) = 0$, we say that at the point $z \in A \times D$ a *resonance correlation* takes place between the frequencies (resonance).

Let us introduce a set of resonance vectors,

$$K_1 = \{k \in K : k\omega(z) = O, \ k \neq 0, \ z = (0, y), \ y \in D\}.$$

The stability conditions for stationary regimes which are given in Ref. [23] hold in the case when slow variables change in the $O(\mu)$-neighbourhood of the stationary position of equilibrium. The combinative frequencies $k\omega$, $k \in K_1$ have in this case the order $O(\mu)$ and therefore, the combinative phases kq, $k \in K_1$ also become slow variables. As a result, the number of slow variables decreases in the $O(\mu)$-neighbourhood.

In Chapter 4 we stated the sufficient conditions for stability of the multi – frequency systems (3.1) with respect to slow variables. These conditions are based on an estimate of the small denominators $k\omega$, $k \in K_1$ in the ϵ-neighbourhood of the equilibrium position, where ϵ is a quantity taken from the definition of Lyapunov stability; this estimate was proposed in Refs. [62] and [44].

We introduce the following notations

$$\Phi_0 = \frac{\partial u}{\partial z} Z_0,$$

$$\Phi_1 = \frac{\partial u}{\partial z} \sum_{k \in K_1} Z_k \exp(\imath kq).$$

Theorem 6.2 *Suppose that:*

a) For the system (3.3) *there exists a Lyapunov function* $v(z)$ *which is positive – definite in the variables* x, *and the function* $\phi_0(z)$ *being negative – definite in the variable* x.

b) There exists a non-negative function $u = u(z)$ *such that for any* $\epsilon > 0$, *($\epsilon < \epsilon_0$) there exist numbers* $\eta' > 0$, $\eta'' > 0$ *($\eta' < \eta'' < \eta_0(\epsilon)$),* $c' > 0$, $c' < c(\epsilon)$ *with*

$$\sup(\Phi_0(z) + \Phi_1(z,q)) < 0, \quad z \in G(\eta',\epsilon) \times D, \quad q \in \mathbf{R}^r,$$

$$\inf_{k,z}|k\omega(z)| > 0, \quad k \in K_1, z \in [H(\eta'',\epsilon) \cap S(c',\epsilon)] \times D.$$

c)

$$\inf_{k,z}|k\omega(z)| > 0, \quad k \in K_2 = K \setminus \{0,k_1\}, z \in Q(\epsilon_0) \times D.$$

d) The functions Z_k, $k \in K$, $\dfrac{dv}{dz}$, $\dfrac{du}{dz}$, ω *satisfy the Lipschitz condition and are bounded (except* ω*) in the region* $z \in Q(\epsilon_0) \times D$.

Then the system (3.1) *is* (x,μ)*-stable.*

Proof. Let us show that for the system (3.1) the condition of Theorem 6.1 follow from the conditions of Theorem 6.2.

It is easy to show that conditions (a) to (c) of Theorem 6.1 are fulfilled. Let us check condition (d). In this case, the function Φ has the form

$$\Phi = \Phi_0(z) + \Phi_1(z, q) + \Phi_2(z, q), \tag{3.5}$$

where

$$\Phi_2 = \frac{\partial u}{\partial z} \sum_{k \in K_2} Z_k \exp(\imath k q),$$

and

$$\phi = \phi_0(z) + \phi_1(z, q) + \phi_2(z, q), \tag{3.6}$$

with

$$\phi_j = \frac{\partial v}{\partial z} \sum_{k \in K_j} Z_k \exp(\imath k q), \quad j = 1, 2.$$

By virtue of conditions (a) and (b) there exist numbers $k > 0$ and $k_1 > 0$ such that

$$\phi_0(z) < -k_0, \quad z \in S(c', \epsilon) \times D, \quad \| \hat{y} \| < \hat{\epsilon}, \tag{3.7}$$

$$\Phi_0(z) + \Phi_1(z, q) < -k_1,$$
$$z \in G(\eta', \epsilon) \times D, \quad q \in \mathbf{R}^r, \quad \| \hat{y} \| < \hat{\epsilon}. \tag{3.8}$$

Further, according to condition (c) and the fact that functions Z_k are bounded, there exists $T_1 > 0$ such that for all $T \geq T_1$

$$\frac{1}{T} \Big| \int_0^T \Phi(z, \omega(z)t + q_0) \, dt \Big| \leq \frac{k_1}{2} \tag{3.9}$$

for all $z \in Q(\epsilon_0) \times D, q \in \mathbf{R}^r, \| \hat{y} \| \leq \hat{\epsilon}$

From (3.5), (3.8) and (3.9) it follows that

$$\sup_{z_0, q_0, t_0} \frac{1}{T} \int_{t_0}^{t_0+T} \Phi(z_0, \omega(z_0)(t - t_0) + q_0) \, dt \leq -\frac{k_1}{2},$$
$$z_0 \in G(\eta', \epsilon) \times D, \quad q_0 \in \mathbf{R}^r, \quad t_0 \geq 0. \tag{3.10}$$

The quantity T in the inequality above satisfies the condition $T \geq T_1$. Since for

$$z \in [H(\eta'', \epsilon) \cap S(c', \epsilon)] \times D, \quad \| \hat{y} \| \leq \hat{\epsilon},$$

by virtue of condition (b) of the theorem, the estimate of the small denominators $|k\omega(z)| \geq \inf_{k,z} |k\omega(z)| > 0$, $k \in K_1$ is valid, then, according to conditions (a), (b), and (d), there exists $T_2 > 0$ such that

$$\frac{1}{T} | \int_0^T [\phi_1(z, \omega(z)t + q) + \phi_2(z, \omega(z)t + q)] \, dt | \leq \frac{k_0}{2} \qquad (3.11)$$

for any $T > T_2$ and any $q \in \mathbf{R}^r$.

Combining (3.7), (3.11), and (3.6), we obtain

$$\sup_{z_0, q_0, t_0} \frac{1}{T} \int_{t_0}^{t_0+T} \phi(z_0, \omega(z_0)(t - t_0) + q_0) \, dt \leq -\frac{k_0}{2},$$

where

$$z_0 \in [H(\eta'', \epsilon) \cap S(c', \epsilon)] \times D, \quad \| \hat{y}_0 \| \leq \hat{\epsilon},$$

$$q_0 \in \mathbf{R}^r, \quad t_0 \geq 0, \quad T \geq T_2.$$

Now, if we set $T = \max\{T_1, T_2\}$, then the inequalities (3.10) and (3.12) remain valid. Therefore, condition (d) of Theorem 6.1 is fulfilled. The theorem is proved.

Remark. If the system (3.1) is (y, μ)-stable, then conditions (b) and (c) can be checked under the additional restriction $\| \hat{y} \| \leq \hat{\epsilon}(\epsilon)$.

6.4   Stability of Gyroscope with No – Contact Suspension

The system of differential equations, which describe in first approxima-tion the motions of a gyroscope with a no-contact suspension on a base vibrating with respect to the mass centre, can be written as follows [80]

$$\dot{\Theta} = \mu[X_{10}(\Theta,l) + X_{11}(\Theta,l,\rho,q)],$$
$$\dot{l} = \mu[X_{20}(\Theta,l) + X_{21}(\Theta,l,\rho,q)],$$
$$\dot{\rho} = \mu Y_{11}(\Theta,l,\rho,q), \qquad\qquad (4.1)$$
$$\dot{\sigma} = \mu Y_{21}(\Theta,l,\rho,q),$$
$$\dot{q} = \chi(\Theta,l) + \mu Q(\Theta,l,\rho,q).$$

Here μ is a small parameter ($\mu = 10^{-8} - 10^{-10}$), the angles ρ and σ determine the position of the vector of kinetic momentum with respect to the reference frame moving together with the centre of mass, l is the value of the kinetic momentum, $q = \gamma - \psi - \phi$ where γ is the angle of the induced harmonic motion of the base of the casing with the frequency ω, the angles ψ , ϕ and Θ are the Euler angles of the rotor with respect to the reference frame, relative to the kinetic momentum vector. We also note that the angle is equal to the angle between the vibration axis of the casing and the kinetic momentum vector. Using the notation of Ref. [80], we write the RHS of the system (4.1):

$$X_{10} = \frac{\sin\Theta}{2l}\left\{-\frac{1}{2}[(1 + \cos\Theta)\mathrm{Im}w(\imath l - \imath\nu)-\right.$$
$$-(1 - \cos\Theta)\mathrm{Im}w(\imath l + \imath\nu) + 2\mathrm{Im}w(\imath\nu)]\cos^2\beta +$$
$$\left.+2\cos\Theta\mathrm{Im}w(\imath l)\sin^2\beta\right\},$$

$$X_{20} = -\frac{1}{4}[(1 - \cos\Theta)^2\mathrm{Im}w(\imath l + \imath\nu) +$$
$$+(1 + \cos\Theta)^2\mathrm{Im}w(\imath l - \imath\nu)]\cos^2\beta - \sin^2\Theta\mathrm{Im}w(\imath l)\sin^2\beta,$$

$$X_{11} = -\frac{\sin\Theta}{2l}(A\sin q + B\cos q)\sin\rho\cos\beta,$$

$$X_{21} = \frac{1 + \cos\Theta}{2}(A\sin q + B\cos q)\sin\rho\cos\beta,$$

$$Y_{11} = \frac{1 + \cos\Theta}{2}(A\sin q + B\cos q)\cos\rho\cos\beta,$$

$$Y_{21} = \frac{1 + \cos \Theta}{2} (A \cos q - B \sin q) \cot \rho \cos \beta,$$

$$Q = \frac{1 + \cos \Theta \cos 2\rho}{2l \sin \rho} (A \cos q - B \sin q) \cos \beta -$$

$$- \frac{1 + \cos \Theta}{2l} \{[(1 + \cos \Theta) \mathrm{Re} w(\imath l - \imath \nu) -$$

$$- (1 - \cos \Theta) \mathrm{Re} w(\imath l + \imath \nu) - 2 \cos \Theta \mathrm{Re} w(\imath \nu)] \cos^2 \beta -$$

$$- 4 \cos \Theta \mathrm{Re} w(\imath \nu) \sin^2 \beta\},$$

$$\chi = \frac{lk \cos \Theta}{1 + k} - l + \omega,$$

$$\chi = \frac{lk \cos \Theta}{1 + k},$$

$$A = \frac{1}{2} \mathrm{Re} v(\imath \omega),$$

$$B = \frac{1}{2} \mathrm{Im} v(\imath \omega).$$

Here $v(D) = K_0 a(D)[K_0 a(D) + D^2 b(D)]^{-1}$, $w(D) = D^2 v(D) a(D)$, $b(D)$ are polynomials with constant coefficients K_0, b are constants, $\imath^2 = -1$, β is the angle between an imbalance vector and the equatorial plane of the central ellipsoid of inertia of the rotor relative to the principal central axes, $k = I_3/I_1 - 1$, where $I_1 = I_2$, I_3 are moments of inertia of the rotor.

In addition to the system (4.1) we consider the system

$$\dot{\Theta} = \mu X_{10}(\Theta, l), \quad \dot{l} = \mu X_{20}(\Theta, l), \quad \dot{\rho} = 0, \quad \dot{\sigma} = 0. \qquad (4.2)$$

which contains zero harmonics only (see Section 3 of the present chapter).

The first two equation of this system have equilibrium position

$$\Theta = \Theta_0 = 0, \quad l = l_0, \qquad (4.3)$$

where l_0 is a root of the equation

$$\mathrm{Im} w(\imath l - \imath \nu) = 0, \quad \nu_0 = \nu(\Theta_0, l_0). \qquad (4.4)$$

When the inequalities

$$\frac{1}{2}\text{Im}w(\imath\nu_0)\cos^2\beta + \text{Im}w(\imath l_0)\sin^2\beta > 0$$

$$\cos^2\beta\frac{\partial}{\partial l}\text{Im}\left(\frac{\imath l}{1+k}\right)|_{l=l_0} > 0 \qquad (4.5)$$

are fulfilled, then, as was shown in Ref. [80], in a course of investigation of a non - resonance gyroscopic system, the equilibrium position of the system (4.2) is asymptotically stable in a part of the variables.

Let us assume that at the point $\Theta = \Theta_0 = 0$, $l = l_0$ we have the equation

$$\chi(\Theta_0, l_0) = 0. \qquad (4.6)$$

Relation (4.6) determines the condition of the resonance between frequencies of the induced, precession and proper motions of the gyroscope.

Note that the RHS of the system (4.1) has at $\rho = 0, \pi$ a singularity of the form $(\sin\rho)^{-1}$, therefore, we consider only the solution which satisfy the inequalities

$$0 < \gamma_1 \leq \rho \leq \gamma_2 < \pi. \qquad (4.7)$$

Theorem 6.3 *In the region* (4.7) *the system* (4.1) *is* (Θ, l, μ)-*stable at the point* (4.3) *with resonance* (4.6).

Proof. We make use of Theorem 6.2. As the function v we take

$$v = (\Theta - \Theta_0)^2 + (l - l_0)^2.$$

By virtue of the inequalities (4.5) there exists $\epsilon_0 > 0$ such that in the ball

$$Q(\epsilon_0) = \{(\Theta, l) : (\Theta - \Theta_0)^2 + (l - l_0)^2 < \epsilon_0^2\}$$

the function $\phi_0 = \dfrac{\partial v}{\partial \Theta}X_{10} + \dfrac{\partial v}{\partial l}X_{20}$ is negative - definite.

As $u(\Theta, l)$ we take the following function

$$u(\Theta, l) = \frac{2l - l_0}{l}\Theta^2,$$

then

$$\Phi = \Theta^2 \frac{l_0}{l^2}(X_{20} + X_{21}) + \frac{2l - l_0}{l}2\Theta(X_{10} + X_{11}).$$

Let us expand the function $\Phi(\Theta, l, q)$ in powers of Θ and $(l - l_0)$. We obtain

$$\Phi = -\frac{\Theta^2}{l_0}[2\mathrm{Im}w(\imath l_0)\sin^2\beta + \mathrm{Im}w(\imath\nu_0)\cos^2\beta] + r(\Theta, l - l_0, q), \quad (4.8)$$

with

$$|r(\Theta, l - l_0, q)| < \Theta^2 O(\epsilon_0). \tag{4.9}$$

It follows from the first inequality in (4.5) combined with (4.8) and (4.9) that there always exists an $\epsilon_0 > 0$ such that the function Φ is negative in the ball $Q(\epsilon)$ for $\Theta \neq 0$. Then, for fixed $\epsilon > 0$ ($\epsilon < \epsilon_0$) we set $\eta' = (\epsilon/3)^2$. Let $c' = (2\epsilon/3)^2$. We construct the sets

$$H(\eta', \epsilon) = \{(\Theta, l) \in Q(\epsilon) : u(\Theta, l) < \eta'\},$$
$$G(\eta', \epsilon) = \{(\Theta, l) \in Q(\epsilon) : u(\Theta, l) = \eta'\},$$
$$P(c', \epsilon) = Q\left(\frac{2}{3}\epsilon\right), \quad S(c', \epsilon) = \partial Q\left(\frac{2}{3}\epsilon\right).$$

Note that since no oscillating terms are present ($K_2 = \emptyset$), one can set $\eta' = \eta''$ in Theorem 6.1. Having expanded the function $\chi(\Theta, l)$ in powers of Θ and $l - l_0$ in a neighbourhood of the point (3.3), we obtain

$$\chi(\Theta, l) = -\frac{l - l_0}{1 + k} + \cdots$$

Terms of the second and higher order in Θ and $l - l_0$ are not written down.

In the region $H(\eta'', \epsilon) \cap S(c', \epsilon)$, the minimal value of $|l - l_0|$ has the order of ϵ and condition (b) of Theorem 6.2 is fulfilled. The other conditions are satisfied as well. The theorem is proved.

Theorem 6.4 *The system* (4.1) *is* (Θ, l, ρ, μ)*-stable on the segment* $0 \le \tau \le \mu^{-1}$ *at the point* $\Theta = \Theta_0$, $l = l_0$, $\rho = \rho_0$, $\gamma_1 < \rho_0 < \gamma_2$ *with resonance* (4.6).

Proof. The system (4.1) gives the identity

$$l(\tau) \cos \rho(\tau) - l(0) \cos \rho(0) =$$
$$= \mu \int_0^\tau X_{20}(\Theta(t), l(t)) \cos \rho(t) \, dt. \tag{4.10}$$

which is obtained by multiplying the second equation by $\cos \rho$ and the third equation by $(-l \sin \rho)$ and by integrating the sum of these equations on the segment $0 \le t \le \tau$.

The function X_{20} is continuous; it becomes zero at the point $\Theta = \Theta_0$, $l = l_0$. Thus, it follows from Eq. (4.10) and Theorem 6.3 that the angle ρ is stable at an arbitrary point from the interval $\gamma_1 < \rho_0 < \gamma_2$ for $0 \le \tau \le \mu^{-1}$. The theorem is proved.

Theorem 6.5 *The system* (4.1) *is* $(\Theta, l, \rho, \sigma, \mu)$*-stable on the segment* $0 \le \tau \le \mu^{-1}$ *at the point* $\Theta = \Theta_0$, $l = l_0$, $\rho = \frac{1}{2}\pi$, $\sigma = \sigma_0$.

In fact,

$$\sigma(\tau) - \sigma(0) = \mu \int_0^\tau \frac{1 + \cos \Theta}{2l} (A \cos q - B \sin q) \cot \rho \cos \beta \, dt.$$

By virtue of Theorem 6.4, the integral expression can be made arbitrarily small in a neighbourhood of the point $\rho_0 = \frac{1}{2}\pi$ on the time interval $0 \le \tau \le \mu^{-1}$, and hence the assertion of Theorem 6.5 follows.

Investigation of stability of the complete equation system which describes resonance motions of the gyroscope on a vibrating base [80] involves additional difficulties due to singularities in the system of equations constructed by using Euler angles. These equations are discussed in Refs. [47], [29], [49], where local variable changes are introduced in order to avoid these singularities. It is established that, for the model described by equation with singularities, the stability conditions remain the same [49]. Note that the conditions obtained above do not contain resonance phases (which, generally speaking, are difficult to observe), but do contain physical parameters that can be measured and controlled.

CHAPTER 7

Averaging and Stability in Systems of Equations with Delay

In the present chapter, we consider averaging problems and stability studies in multi – frequency systems of differential equations with a deviating argument. We propose new schemes for averaging multi – frequency systems with a variable deviation of the argument on an asymptotically large time interval. The stability of isolated points of resonance in a multi – frequency system with deviating argument are investigated. Cases are selected in which introduction of delay of a variable leads to changes in the averaged system, causes new resonances, and can change the stability type of an isolated point of resonance in the corresponding system of ordinary differential equation.

The results are applied to study evolution in a compact system of two celestial bodies.

187

7.1 Averaging in Systems with Delay

Here, we consider averaging problems arising in multi – frequency systems of standard-form differential equation with deviating argument. We also investigate stability of isolated points of resonance in such systems, and stability of systems containing delay with respect to perturbations under assumption that the unperturbed system has a stable position of equilibrium [69], [70], [71], [50].

To begin with, we examine basic facts concerning averaging in multi – frequency systems. Using the methods presented in Chapter 1, we justify averaging in systems with delay. Here one should distinguish between two situations: when the delay changes the system to be averaged, and when the result of averaging is the same as in the case without delay.

1. We consider a system that contain delay only in slow variables:

$$\dot{x} = \mu X[x(t), x(t - \tau(x, \psi, t)), \psi(t)],$$

$$\dot{\psi} = \omega[x(t), x(t - \tau(x, \psi, t))] + \mu \Phi[x(t), x(t - \tau(x, \psi, t)), \psi(t)],$$

$$x \in D \subset \mathbf{R}^n, \quad t > 0, \quad \dim \psi = m \geq 1, \quad 0 \leq \mu \ll 1. \tag{1.1}$$

The corresponding averaged system has the form

$$\dot{\xi} = \mu X_0(\xi, \xi),$$

where

$$X_0(x, y) = \frac{1}{(2\pi)^m} \int_0^{2\pi} X(x, y, \psi) \, d\psi. \tag{1.2}$$

In the ϵ-neighbourhood of the solution $\xi = \xi(\mu t)$ of the averaged system (1.2) we represent the RHS of (1.1) as

$$X(x, y, \psi) = X_0(x, y) + X_\epsilon(x, y, \psi) + \tilde{X}(x, y, \psi). \tag{1.3}$$

The term $X_\epsilon(x, y, \psi)$ contains only those terms of the Fourier series of the function $X(x, y, \psi)$ for which $k\omega(x, y)| < \sigma(\epsilon)$, while $\tilde{X}(x, y, \psi)$ contains the oscillating terms.

We calculate the minimum of the derivative of the resonance frequency of the corresponding system without delay in the direction of the vector $\bar{X} = X_0 + X_\epsilon$ along the curve $\xi(\mu t)$ with respect both to $\psi \in [0, 2\pi]^m$ and all k present in $X_\epsilon(x, y, \psi)$;

$$a(\epsilon) = \min \| \frac{\partial k \omega(x, x)}{\partial x} \bar{X}(x, x, \psi) \| . \tag{1.4}$$

Theorem 7.1 . *Suppose that in the domain D for $t > 0$ the following conditions are fulfilled:*

a) The RHS of the system (1.1) and (1.2) satisfy conditions of the theorem on the existence and uniqueness of solutions.

(b) $X(x, y, \psi)$, $\Phi(x, y, \psi)$ are periodic in ψ with the period 2π and the functions X and Φ can be expanded into absolutely and uniformly convergent Fourier series.

c) $\| X(x, y, \psi) \| < M$, $\| \Phi(x, y, \psi) \| < M$, $|\omega(x, y)| < M$. $|\omega'_x(x, y)| < \omega'_0$, $\| \omega'_y(x, y) \| < \omega'_0$ with $X(x, y, \psi)$ satisfying the Lipschitz condition in x and y with the Lipschitz constant λ.

(d) For each ϵ, $0 < \epsilon < \epsilon_0$, there exist $\sigma(\epsilon)$ and $a(\epsilon)$ such that, in accordance with representation (1.3) for the RHS of the initial system, $a(\epsilon_0) > 0$, where $a(\epsilon)$ is defined by (1.4).

Then for any ϵ, $0 < \epsilon < \epsilon_0$, there exists $\mu(\epsilon)$ such that for all $0 < \mu < \mu_0(\epsilon)$, $0 < t < L/\mu$ (where L is an arbitrary positive number),

$$\| x(t) - \xi(\mu t) \| < \epsilon$$

The proof follows the line of reasoning used in the proof of Theorem 1.4, i.e. by comparing the solutions of the complete and averaged systems and taking into account estimates related to the initial set.

2. Now we consider a system of the form (1.1) containing delay in the fast variables as well:

$$\begin{aligned}
\dot{x} &= \mu X[x(t), x(t - \tau(x, q)), \psi(t), \psi(t - \tau(x, q))], \\
\dot{\psi} &= \omega[x(t), x(t - \tau(x, q))] + \\
&\quad + \mu \Phi[x(t), x(t - \tau(x, q)), \psi(t), \psi(t - \tau(x, q))].
\end{aligned} \tag{1.5}$$

We assume the delay periodically depends on the phases q (dim $q = m_1 \geq 1$):

$$\tau(x,q) = \sum_k \tau_k(x)e^{\imath kq}, \qquad \dot{q} = \nu(x) + \mu\Theta(x,q). \qquad (1.6)$$

Then, with accuracy $O(\mu^2)$, the system (1.5) is transformed into the form

$$\dot{x} = \mu Y[x(t), x(t - \tau(x,q)), \psi(t), q(t)],$$
$$\dot{\psi} = \omega[x(t), x(t - \tau(x,q))] + \mu\Psi[x(t), x(t - \tau(x,q)), \psi, q], (1.7)$$
$$\dot{q} = \nu(x) + \mu\Theta(x,q).$$

where the RHS $X(x, x_t, \psi, \psi_t)$ are represented by the Fourier series

$$X(x, x_t, \psi, \psi_t) = \sum_{k,l} X_{k,l}(x, x_t) \exp \imath(k\psi + l\psi_t) =$$
$$= \sum_{k,l} X_{k,l}(x, x_t) \exp \imath[(k + l)\psi - l\omega(x,x)\tau(x,q)]] =$$
$$= \sum_{k_1,k_2} Y_{k_1,k_2}(x, x_t) \exp \imath(k_1\psi + k_2q) = Y(x, x_t, \psi, q) \qquad (1.8)$$
$$Y_{k_1,k_2}(x, x_t) = \frac{1}{(2\pi)^{m+m_1}} \sum_{k,l} X_{k,l} \int_0^{2\pi} \exp \imath[(k + l - k_1)\psi -$$
$$- \imath(l\omega(x,x)\tau(x,q) + k_2q)] \, d\psi dq.$$

with accuracy $O(\mu)$.

We introduce the notation

$$Y_0(x,y) = \frac{1}{(2\pi)^{m+m_1}} \int_0^{2\pi} Y(x,y,\psi,q) \, d\psi dq. \qquad (1.9)$$

The system

$$\dot{\xi} = \mu Y_0(\xi, \xi) \qquad (1.10)$$

is called averaged with respect to the initial system (1.5).

Thus, the system (1.5) is a multi – frequency system, and its RHS contains both frequencies and phases of the system, as well as frequencies and phases of the delay; the RHS of the averaged system

(1.10) is obtained with the delay phases averaged as well. The sufficient conditions for proximity of the solution of the complete and averaged systems are obtained, as in Chapter 1, by singling out resonance harmonics in the σ-neighbourhood of the solution of averaged system. Then $\epsilon > 0$ and $\sigma(\epsilon) > 0$ are fixed, and the combinative frequencies $k_1\omega(x, x) + k_2\nu(x)$ are decomposed into resonance $(|k_1\omega + k_2\nu| < \sigma(\epsilon))$ and non – resonance $(|k_1\omega + k_2\nu| \geq \sigma(\epsilon))$ frequencies; the RHS $X(x, x, \psi, \psi_t)$ are decomposed correspondingly:

$$
\begin{aligned}
X(x, x, \psi, \psi_t) = Y(x, x, \psi, q) = \\
= Y_0(x, x) + Y_\epsilon(x, x, \psi, q) + \tilde{Y}(x, x, \psi, q) = \\
= \bar{Y}(x, x, \psi, q) + \tilde{Y}(x, x, \psi, q),
\end{aligned} \tag{1.11}
$$

where $Y_\epsilon(x, x, \psi, q)$ denotes the sum of those terms of the Fourier series for which $|k_1\omega(x, x) + k_2\nu(x)| < \delta(\epsilon)$ in the ϵ-neighbourhood of the curve $\xi = \xi(\mu t)$.

As above, we calculate the minimum of the absolute value of derivative of the resonance frequency $k_1\omega(x, x) + k_2\nu(x)$ in the direction of the vector $\bar{Y}(x, x, \psi, q) = Y_0(x, x) + Y_\epsilon(x, x, \psi, q)$. The minimum is taken over all k_1, k_2 (entering $Y_\epsilon(x, x, \psi, q)$), $\psi \in [0, 2\pi]^m$, $q \in [0, 2\pi]^{m_1}$, and x along the curve $\xi = \xi(\mu t)$:

$$
a = \min \left| \frac{\partial(k_1\omega(x, x) + k_2\nu(x))}{\partial x} \bar{Y}(x, x, \psi, q) \right|. \tag{1.12}
$$

In formulation of the theorem on proximity of the complete and averaged systems, we will regard a as strictly positive.

Theorem 7.2 *Let conditions (a) and (c) of Theorem 7.1 be fulfilled. Let the functions $X(x, x_t, \psi, q)$ and $\Phi(x, x_t, \psi, q)$ be expanded in absolutely and uniformly convergent Fourier series in the phases ψ and q. Let us also assume that for any ϵ, $0 < \epsilon < \epsilon_0$ there exist $\sigma(\epsilon)$ and $a(\epsilon)$ such that $a(\epsilon_0) > 0$ ($a(\epsilon_0)$ is defined by Eq. (1.12)).*

Then for any ϵ, $0 < \epsilon < \epsilon_0$ *there exists* $\mu_0(\epsilon)$ *such that for all* μ, t, $0 < \mu < \mu(\epsilon)$, $0 < t < L/\mu$ *(where* L *is an arbitrary positive number), the inequality*

$$\| \, x(t) - \xi(\mu t) \, \| < \epsilon$$

holds.

Note that the averaging is carried out over all the phases, including the delay phase of (1.6). The coefficients in the Fourier series must be written analytically if the RHS of the system (1.7) can be expanded into series in powers of the delay.

7.2   Stability of Systems with Deviating Argument

1. We consider a system of differential equations with delayed argument

$$\dot{x} = f[x(t), x(t - \mu\tau(x,t)), t] + \mu R[x(t), x(t - \mu\tau(x,t)), t], \quad (2.1)$$

where μ is a small parameter, $\dim x = n$, $R(x, x_t, t)$ is the perturbation vector, $\tau(x,t)$ is the delay function, $0 < \tau(x,t) < \tau_0$. The system (2.1) will be examined in the region

$$\| \, x \, \| < H, \quad \| \, x_t \, \| < H, \quad t \geq 0. \quad (2.2)$$

For $t \in E_0 = [-2\tau_0, 0]$, let $x(t) = \kappa(t) \in C^1(E_0)$. Taking $\mu = 0$, we obtain a unperturbed system, which contains no delay and is a system of ordinary differential equations

$$\dot{x} = F(x, t) = f(x, x, t). \quad (2.3)$$

Let the unperturbed system (2.3) have a stable equilibrium position $x = 0$, $F(0, t) = 0$; we investigate the point $x = 0$ in order to verify its (x, μ)-stability with respect to the perturbations R. In this case, the effect of the delay on stability can be quite significant.

The investigation of stability will be carried out using the technique of the perturbed Lyapunov function $v = v_0 + \mu u$, where v_0 is the Lyapunov function and u is determined from the equation

$$\frac{\partial u}{\partial t} + \frac{\partial u}{\partial x}F(x,t) = -\frac{\partial v_0}{\partial x}R(x,x,t). \tag{2.4}$$

After expanding in powers of the small parameter, the derivative of the generalized Lyapunov function takes the form

$$\frac{dv}{dt} = \frac{\partial v_0}{\partial t} + \frac{\partial v_0}{\partial x}f(x,x,t) -$$

$$-\mu\frac{\partial v_0}{\partial x}\tau(x,t)f(x,x,t)f'_y(x,y,t)|_{y=x} + \mu^2 G = \tag{2.5}$$

$$= \frac{\partial v_0}{\partial t} + \frac{\partial v_0}{\partial x}F(x,t) + \mu\phi_\tau(x,t) + \mu^2 G,$$

where G denotes the remainder term of the expansion.

The theorems on (x,μ)-stability for systems with delayed argument are formulated as the theorems in Chapter 2, the only difference being the presence of delay. In the proof one makes use of the theorem on existence and uniqueness for equations with deviating argument [21], [41].

Theorem 7.3 *Suppose that the following conditions are fulfilled:*

a) The RHS of the systems (2.1) and (2.3) satisfy the conditions for existence and uniqueness of a solution.

b) $f(x,y,t)$ satisfies the Lipschitz conditions in the variables x and y.

c) There exist a summable function $M(t)$ and a constant M_0 such that on any finite segment

$$\| R(x,y,t) \| \leq M(t), \qquad \int_{t_1}^{t_2} M(t)\,dt \leq M_0(t_1 - t_2).$$

d) There exists a positive – definite Lyapunov function $v_0(x,t)$ for the system (2.3) which admits an infinitely small upper bound u in x, *has continuous derivatives with respect to both arguments, and satisfies the conditions* $v_0(0,t) = 0$, $\dfrac{dv_0}{dt} \leq 0$.

e) For any $\epsilon > 0$ ($\epsilon < \epsilon_0 < H$) there exist σ_0, γ related by v_0 such that for all ρ, $\gamma \leq \rho \leq \epsilon$ there exists a bounded solution of Eq. (2.4) in the annulus $\rho \leq\parallel x \parallel\leq \epsilon$ for $t > 0$.

f) There exist $\delta > 0$, $l > 0$ such that if $\rho \leq\parallel x \parallel\leq \epsilon$, for $t_0 \leq t \leq t_0 + T$, $T > l$, then, uniformly with respect to $t_0 > 0$

$$\int_{t_0}^{t_0+T} \phi_\tau(\bar{x}(t),t)\,dt \leq -\delta T,$$

where $\bar{x} = \bar{x}(t)$ is the solution of the unperturbed system (2.3) and ϕ_τ is defined by relation (2.5).

g) There exist a summable function $Q(t)$, a constant Q_0, and a nondecreasing function $\chi_1(\alpha)$, $\lim\limits_{\alpha \to 0}\chi_1(\alpha) = 0$, such that for $\rho \leq\parallel x \parallel\leq \epsilon$, $t > 0$

$$\parallel \phi_\tau(x',t,\mu,\epsilon) - \phi_\tau(x'',t,\mu,\epsilon) \parallel\leq \chi_1(\parallel x' - x'' \parallel)Q(t),$$

with $\int_{t_1}^{t_2} Q(t)\,dt \leq Q_0(t_2 - t_1)$ on any finite segment $[t_1,t_2]$.

Then, for any $\epsilon > 0$ there exist $\eta(\epsilon) > 0$, $\mu_0(\epsilon) > 0$ such that for all μ, $0 < \mu < \mu_0(\epsilon)$, all the solution satisfying the inequality $\parallel x(t) \parallel< \eta(\epsilon)$ for $t \in E_0$, satisfy the inequality $\parallel x(t) \parallel< \epsilon$ for all $t > 0$, i.e. the equilibrium position is (x,μ)-stable with respect to the perturbations μR.

The proof follows the scheme employed in Chapter 2 with the only change that presence of the initial set is taken into account.

2. Let the unperturbed system (2.3) be autonomous,

$$\dot{\bar{x}} = f(\bar{x},\bar{x}) \tag{2.6}$$

and

$$\nabla v_0(x)f(x,x) \equiv 0. \tag{2.7}$$

In this case, the main condition (f) of the theorem is not fulfilled, and stability of the equilibrium position of the system must be investigated using the perturbed Lyapunov function of the form

$$v(x,t) = v_0(x) + \mu u_1(x,t) + \mu^2 u_2(x,t), \tag{2.8}$$

where the functions $u_1(x,t)$ and $u_2(x,t)$ are defined as bounded solutions of the system

$$\frac{\partial u_1}{\partial t} + \frac{\partial u_1}{\partial x} f(x,x) = -\nabla v_0(x) R(x,x,t), \tag{2.9}$$

$$\frac{\partial u_2}{\partial t} + \frac{\partial u_2}{\partial x} f(x,x) = -\nabla u_1(x) R(x,x,t). \tag{2.10}$$

Taking (2.7) into account, we obtain

$$\frac{\partial v}{\partial t} = \mu^2 \bar{\phi}_\tau(x,t) + \mu^3 \bar{G}, \tag{2.11}$$

where

$$\bar{\phi}_\tau(x,t) = -\tau(x,t) \frac{\partial f}{\partial y}\Big|_{y=x} \{\nabla v_0(x) R(x,x,t) + \frac{\partial u_1}{\partial x} f(x,x)\}. \tag{2.12}$$

Here, the remainder term $\bar{G}$ contains the functions $f(x,y,t)$, $R(x,y,t)$, $\tau(x,t)$ and their first and second derivatives.

We consider the system with a finite delay

$$\dot{x}(t) = f(x,t) + \mu R[x(t), x(t - \tau(x,t)), t] \tag{2.13}$$

and the unperturbed system

$$\dot{\bar{x}} = f(\bar{x},t). \tag{2.14}$$

As above, to investigate stability we use the perturbed Lyapunov function

$$v(x,t,,\mu,\epsilon) = v_0(x,t) + \mu u(x,t), \tag{2.15}$$

$$\frac{dv}{dt} = \frac{dv_0}{dt} + \mu \frac{\partial v_0}{\partial t} R_\tau(x,t) +$$
$$+\mu \left[\frac{\partial u}{\partial t} + \frac{\partial u}{\partial x} f(x,t) + \frac{\partial v_0}{\partial x} R(x,x,t) \right] + \tag{2.16}$$
$$+\mu \frac{\partial v_0}{\partial t} R_z + \mu^2 G,$$

where the derivatives $\dfrac{du}{dt}$ and $\dfrac{dv_0}{dt}$ are calculated with the help of Eqs. (2.13) and (2.14), respectively. The expansion of $R_z(x,t)$ characterizes the perturbation which appears in the system with a delay:

$$R_\tau(x,t) = \sum_1^{m-1} \frac{h^k(x,t)}{k!} \frac{\partial^k R(x,y,t)}{\partial y^k}\Big|_{y=x}. \tag{2.17}$$

Here $h^k = (h_1^k, \ldots, h_n^k)$,

$$h(x,t) = \sum_1^m \frac{(-1)^k \tau^k(x,t)}{k!} g_k(x,t), \tag{2.18}$$

$\{g_k(x,t)\}$ being a recurrent sequence:

$$g_0(x) = x,$$

$$\ldots$$

$$g_k(x,t) = \frac{\partial g_{k-1}(x,t)}{\partial t} + \frac{\partial g_{k-1}(x,t)}{\partial x}\gamma(x,t), \quad k = \overline{1,\,m+1}.$$

where $\gamma(x,t)$ is defined in terms of known functions.

In the expression (2.16), R_z is a remainder of expansion in the Taylor series in powers of $\tau(x,t)$, G is a remainder of expansion in powers of the small parameter. The perturbation $u(x,t)$ is defined from equation of the form (2.4).

We introduce the notation

$$\phi_\tau(x,t) = \frac{\partial v_0(x,t)}{\partial x} R_\tau(x,t), \tag{2.19}$$

$$\phi_z(x,t) = \frac{\partial v_0(x,t)}{\partial x} R_z(x,t), \tag{2.20}$$

$$\bar{\phi}_\tau(x_0,t_0) = \lim_{T\to\infty} \frac{1}{T}\int_{t_0}^{t_0+T} \phi_\tau(\bar{x},t)\,dt, \tag{2.21}$$

where $\bar{x} = \bar{x}(t)$ is the solution of the system (2.19) emanating from the point (x_0,t_0).

Let us define the perturbation of the Lyapunov function $u(x,t)$ as a solution of the equation

$$\frac{\partial u}{\partial t} + \frac{\partial u}{\partial x}f(x,t) = -\frac{\partial v_0}{\partial x}R(x,x,t), \tag{2.22}$$

bounded in the annulus $0 < \rho < \| x \| < \epsilon$.

We obtain

$$\frac{dv}{dt} = \frac{dv_0}{dt} + \mu\phi_\tau(x,t) + \mu\phi_z(x,t) + \mu^2 G. \tag{2.23}$$

Theorem 7.4 *Suppose that for the system* (2.13) *and* (2.14) *condition* (a) *to* (g) *of Theorem 7.3 are fulfilled. Moreover, let*

$$|\phi_z(x(t), t, x(t_1), x(t_2, t_1, t_2))| < \frac{1}{5}|\bar\phi_\tau(x_0, t_0)|,$$

where $x(t)$ is the integral curve emanating from the point (x_0, t_0), with t_1 and t_2 being some intermediate points.

Then the equilibrium position $x = 0$ of the system (2.14) *is (x, μ)-stable with respect to the perturbation μR.*

Theorems on instability are formulated using the Chetaev function for the unperturbed system and by reversing the sign in the inequality in condition (e).

7.3 Stability in Multi – Frequency Systems with Delay

1. We now turn to investigation of stability of isolated point of resonance in a multi-frequency system in the standard form of the delay type. We can single out as a somewhat simpler situation the case in which delay is only present in slow variables; the situation becomes more complex when delay is present in fast variable as well and is characterized by its own frequency set. In the latter case, due to delay, the number of frequencies in the system increases and resonance phenomena appear due to the interaction of the main system with delayed frequencies.

We consider the system

$$\dot{x} = \mu X[x(t), x(t - \tau(x, \psi)), \psi(t)], \tag{3.1}$$
$$\dot{\psi} = \omega[x(t), x(t - \tau(x, \psi))] + \mu\Phi[x(t), x(t - \tau(x, \psi)), \psi(t)],$$

where $x(t) = \kappa(t)$ is continuous on the initial set $[-2\tau_0, 0] = E_0$, $\dim x = n$, $\dim \psi = \dim \omega = m \geq 2$, $0 < \tau(x, \psi) < \tau_0$.

We assume that the system (3.1) has a resonance point x_0 in the region

$$\| x - x_0 \| < H, \quad \psi \in [0, 2\pi]^m. \tag{3.2}$$

We also examine the averaged system

$$\dot{\xi} = \mu X(\xi, \xi), \quad t > 0, \quad \xi(t) = \kappa(t), \quad t \in E_0 \tag{3.3}$$

which has the stable equilibrium position x_0, $X(x_0, x_0) = 0$. For the system (3.3) there exists a Lyapunov function $v_0(x)$ such that

$$\nabla v_0(x) X_0(x, x) \leq 0.$$

We investigate stability of the system (3.1) by making use of the Lyapunov function

$$v(x, \psi, \mu, \epsilon) = v_0(x) + \mu U(x, \psi) \tag{3.4}$$

in the annulus $0 < \rho < |x - x_0| < \epsilon$. We obtain an expression for derivative of v by expanding this function in powers of μ and $\tau(x, \psi)$ for $t \geq 2\tau_0$:

$$\frac{dv}{dt} = \mu \nabla v_0(x) X(x, x, t) + \mu \frac{\partial U}{\partial \psi} \omega(x, x) +$$

$$+ \mu^2 \nabla v_0(x) X^\tau(x, t) + \mu^2 \frac{\partial U}{\partial \psi} [\omega^\tau(x, \psi) + \Phi(x, x, \psi)] + \tag{3.5}$$

$$+ \mu^2 \frac{\partial U}{\partial x} X(x, x, \psi) + \mu^2 \frac{\partial U}{\partial \psi} \omega_z + \mu^3 G.$$

Here

$$X^\tau(x, \psi) = \frac{\partial X(x, y, \psi)}{\partial y} \Big|_{y=x} h(x, \psi),$$

$$\omega^\tau(x, \psi) = \frac{\partial o(x, y)}{\partial y} \Big|_{y=x} h(x, \psi) \tag{3.6}$$

characterizes effect of the delay, with

$$h(x,\psi) = \sum_{1}^{p} \frac{(-1)^k \tau^k(x,\psi)}{k!} \frac{\partial^{k-1} X(x,x,\psi)}{\partial \psi^{k-1}} \omega^{k-1}(x,x).$$

The functions ω_z and G are the remainder terms of the expansions.

We now make use of the definition of resonance (Chapter 4, Section 1) and decompose the harmonics as before (Chapter 4, Section 3), according to the magnitude of their combinative frequencies:

$$X(x,x,\psi) = X_0(x,x) + X_\epsilon(x,x,\psi) + \tilde{X}(x,x,\psi),$$

$$(3.7)$$

$$Y^\tau(x,\psi) = Y_0^\tau(x) + Y_\epsilon^\tau(x,\psi) + \tilde{Y}^\tau(x,\psi),$$

where

$$Y^\tau(x,\psi) = \nabla v_0(x) X^\tau(x,\psi) + \frac{\partial U}{\partial \psi}[\omega^\tau(x,\psi) +$$

$$+\Phi(x,x,\psi)] + \frac{\partial U}{\partial x} X(x,x,\psi). \qquad (3.8)$$

The function U is defined by the equation

$$\frac{\partial U}{\partial \psi}\omega(x,x) + \nabla v_0(x)\tilde{X}(x,x,\psi) = 0. \qquad (3.9)$$

Let us introduce the notation

$$\phi_\tau(x,\psi) = \nabla v_0(x) X_\epsilon(x,x,\psi) + \mu \nabla v_0(x) Y_0^\tau +$$

$$+\mu Y_\epsilon^\tau(x,\psi) + \mu^2 \tilde{Y}^\tau(x,\psi), \qquad (3.10)$$

$$R_\epsilon = \max_{\|x-x_0\|<\epsilon} \| \nabla v_0(x) X_\epsilon(x,x,\psi) \|$$

Theorem 7.5 *Suppose that the following conditions are fulfilled:*

a) The RHS of the system (3.1) satisfy the conditions for existence and uniqueness of solutions.

b) The functions $X(x,y,\psi)$, $\Phi(x,y,\psi)$ are periodic in ψ with the period 2π and can be expanded into absolutely and uniformly convergent Fourier series in the region $\| x - x_0 \| < H$, $\| y - y_0 \| < H$.

c) The function $X(x, y, \psi)$ is p times continuously differentiable with respect to ψ; the continuous derivatives ω'_y, ω''_{yy}, X'_y, X''_{yy}, Φ'_y do exist.

d) The resonance at the point x_0 is r_0-isolated.

e) The continuous function $\tau(x, \psi)$, periodic in ψ with the period 2π can be expanded into absolutely and uniform convergent Fourier series.

f) For the averaged system (3.3) there exists a positive–definite Lyapunov function having an infinitely small upper bound.

g) In the annulus $0 < \rho < |x - x_0| < \epsilon$, $\psi \in [0, 2\pi]^m$, $t > 2\tau_0$, there exist a solution U of Eq. (3.9) and a constant U_0 such that

$$|U(x, \psi)| < U_0.$$

h) There exist a summable function $Q(t)$, a constant Q_0 and a non-decreasing function $\chi_1(\alpha)$, $\lim_{\alpha \to 0}\chi_1(\alpha) = 0$, such that in the region $\| x - x_0 \| \leq H$, $t \geq 2\tau_0$

$$\| \phi_\tau(x', \psi) - \phi_\tau(x'', \psi) \| \leq \chi_1(\| x' - x'' \|)Q(t),$$

with $\int_{t_1}^{t_2} Q(t)\,dt \leq Q_0(t_2 - t_1)$ on any finite segment $[t_1, t_2]$.

i) The function $Y_0^\tau(x)$ is negative – definite and

$$\max_{\gamma(\epsilon) < |x - x_0| < \epsilon} |Y_\epsilon^\tau(x, \psi)| < \frac{1}{6} \max_{\gamma(\epsilon) < |x - x_0| < \epsilon} [-Y_0^\tau(x)].$$

j) The remainders R_ϵ and ϕ_x are sufficiently small.

Then the point $x = x_0$ of the system (3.1) is (x, μ)-stable for $0 < \mu_1(\epsilon) < \mu < \mu_0(\epsilon)$.

Remark 1. If the RHS of the system (3.1) are trigonometric polynomials in ψ, then $X_\epsilon(x, x, \psi) \equiv 0$, $R_\epsilon \equiv 0$, $\mu_1(\epsilon) = 0$.

Remark 2. The (x, μ)-stability of the resonance point occurs by virtue of the properties of the averaged system, when the main condition of stability is the asymptotic stability of the equilibrium position of the averaged system.

2. Let us now consider a system that contains delay both in slow and fast variables, provided that the time dependence of the delay is close to almost periodic, and thus the delay substantially affects stability:

$$\dot{x} = \mu X[x(t), x(t - \tau(x,q)), \psi(t), \psi(t - \tau(x,q))],$$
$$\dot{\psi} = \omega[x(t), x(t - \tau(x,q))] + \tag{3.11}$$
$$\mu\Phi[x(t), x(t - \tau(x,q)), \psi(t), \psi(t - \tau(x,q))],$$

where $\tau(x,q) = \sum_j \tau_j(x)e^{ijq}$

$$\dot{q} = \nu(x) + \mu\theta(x,q),$$

dim$\psi = m$, dim$q = m_1$.

We introduce the averaged system

$$\dot{\xi} = \mu Y_0(\xi, \xi)$$

by integrating over the phases ψ and q:

$$Y_0(x,y) = \frac{1}{(2\pi)^{m+m_1}} \int_0^{2\pi} Y(x,y,\psi,q)\, d\psi dq. \tag{3.12}$$

We assume that: *a) The system has an isolated point x_0 (in the sense of the definition in Chapter 4) which is the equilibrium position of the averaged system (3.12)*

$$Y_0(x_0, x_0) = 0, \quad |x - x_0| < H, \quad \psi \in [0, 2\pi]^m, \quad q \in [0, 2\pi]^m.$$

b) the RHS of the system (3.11) satisfy the conditions of the existence and uniqueness of solutions.

c) The functions $X(x, \psi, q)$, $\Phi(x, \psi, q)$ are bounded and periodic in ψ, q with the period 2π and can be expanded in these variables into a Fourier series which is absolutely and uniformly convergent; the series for $\tau(x,q)$ converges absolutely and uniformly as well.

d) For the averaged system there exists a positive − definite Lyapunov function $v_0(x)$ ($v_0(x_0) = 0$) for which $\nabla v_0(x)X_0(x) \leq 0$ when $\| x - x_0 \| < H$.

If the non-oscillating part of the derivative of $v_0(x)$ along the system where $\tau = 0$ tends to 0 as $\epsilon \to 0$, then the above-stated conditions are sufficient for stability on a finite time interval (see Chapter 4, Section 4).

e) There exists a monotonically increasing function $\Omega(\epsilon)$ ($\Omega(0) = 0$) and a representation

$$X(x, \psi, \psi) = X_0(x) + X_\epsilon(x, \psi) + \tilde{X}(x, \psi) \tag{3.13}$$

such that

e') The term $X_\epsilon(x, \psi)$ contains those terms of the Fourier series for the function $X(x, \psi, \psi)$ whose frequencies in the ϵ-neighbourhood of the point x_0 satisfy the inequality $|k\omega(x)| < \Omega(\epsilon)$;

e") The term $\tilde{X}(x, \psi)$ contains those terms of the Fourier series which satisfy the inequality $|k\omega(x)| > \Omega(\epsilon)$ and those that correspond to the vector k_0 and its multiples.

We investigate stability by using the perturbed Lyapunov function

$$v(x, \psi, \phi) = v_0(x) + \mu u_1(x, \psi) + \mu u_2(x, \psi, \phi)$$

where u_1 and u_2 are defined by the equations

$$\frac{\partial u_1}{\partial \psi} \omega(x) + \nabla v_0 \tilde{X} = 0,$$

$$\tag{3.14}$$

$$\frac{\partial u_2}{\partial \psi} \omega(x) + \frac{\partial u_2}{\partial \phi} \nu(x) + \nabla v_0 \tilde{Y}(x, \psi, \phi) = 0.$$

Let us differentiate $v(x, \psi, \phi)$ and use the initial system, taking into consideration (3.13) and (3.14). By expanding the derivative in powers of the delay, we obtain

$$\frac{dv}{dt} = \mu \nabla v_0 X_0(x) + \mu \nabla v_0 Y_0(x) +$$
$$+ \mu \nabla v_0 [X_\epsilon(x, \psi) + Y_\epsilon(x, \psi) + z] + \mu^2 R. \tag{3.15}$$

Here we used the functions

$$Y(x, \psi, q) = \sum_1^{h-1} \frac{(-1)^j (\omega \tau)^j}{j!} \frac{\partial^j X(x, \psi, q)}{\partial q^j} \Big|_{q=\psi} \qquad (3.16)$$

and their representation similar to (3.13)

$$Y(x, \psi, q) = Y_0 + Y_\epsilon(x, \psi, q) + \tilde{Y}(x, \psi, q).$$

We also write the expressions for the remainder term of the series (3.16):

$$z = \frac{(-1)^h (\omega \tau)^h}{h!} \frac{\partial^h X(x, \psi, q)}{\partial q^h} \Big|_{q=\tilde{\psi}}, \quad \tilde{\psi} \in [0, 2\pi]^m.$$

The remainder term R of the series (3.17) contains the functions $X(x, \psi, q)$, $\Phi(x, \psi, q)$, $v_0(x)$, $u_1(x, \psi)$, $u_2(x, \psi)$ and their derivatives.

Theorem 7.6 *Suppose that conditions (a) - (e) are fulfilled, together with the following condition :*

(f) For $\| x - x_0 \| < H$ the function $\nabla v_0 Y_0(x)$ is positive-definite. Moreover,

g) There exists a non − decreasing function $\chi(\alpha)$, $\lim\limits_{\alpha \to 0} \chi(\alpha) = 0$, such that for $\| x - x_0 \| \le H$, $q \in [0, 2\pi]^{m_1}$, $\psi \in [0, 2\pi]^m$

$$|\nabla v_0(x') X(x', \psi, q) - \nabla v_0(x'') X(x'', \psi, q)| \le \chi(|x' - x''|).$$

h) There exist functions $\gamma(\epsilon)$, $0 < \gamma(\epsilon) < \gamma_0(\epsilon)$ such that for $\gamma(\epsilon) < \| x - x_0 \| < \epsilon$ the inequality

$$\max_{|x-x_0| \le \epsilon} |\nabla v_0 [X_\epsilon(x, \psi) + Y_\epsilon(x, \psi, q) + z]| < -\frac{1}{2} \nabla v_0 Y_0(x)$$

holds.

i) For any $\rho \ge \gamma(\epsilon)$, $\rho < \epsilon$ there exist bounded solutions of Eq. 3.14) in the annulus $\rho < \| x - x_0 \| < \epsilon$ and a constant R_0 such that $R| < R_0$.

Then there exist $\eta(\epsilon)$ and $\mu_0(\epsilon)$ such that any solution $x = x(t)$, $\psi = \psi(t)$ of the system (3.3), which for $t \leq 0$ satisfies in x the condition $\| f(t) - x_0 \| < \eta(\epsilon)$, satisfies in x the inequality

$$\| x(t) - x_0 \| < \epsilon$$

for all $t > 0$ and $\mu < \mu_0(\epsilon)$.

The proof is based on Theorem 2.1.

7.4 Effect of Variable Tide Delay on the Evolution of Orbital Elements of a Tide - Forming Body

1. The theory of lunar tides [36], [37], [20], [74] takes into account the effect of tide delay with respect to the lunar phase, which in this case is considered to be constant. Tide delay is caused by rotation of the Earth and the friction related to this rotation and causing the tidal bulge to deviate from the Moon. The evolution of the parameters of the Moon's orbit under the influence of the delayed tide is described by Gauss' equations [74] which are in this case equations with deviating argument (here the argument is a phase of the elliptic motion).

Let us write down only one equation from this system, namely the one for the semi-axis a of the lunar orbit:

$$\frac{da}{dt} = \frac{2}{n\sqrt{1 - e^2}} \left(Re \sin f + \frac{a(1 - e^2)}{r} S \right). \qquad (4.1)$$

Here, n is the mean motion; e is the eccentricity; r is the radius vector; f is the actual anomaly of the Moon; R is the force component acting on the Moon in the direction of the radius vector; S is the force component perpendicular to R and lying in the orbital plane. The values of R and S are determined by the potential due to tidal friction and having the form

$$R = -\frac{3}{2}\frac{\gamma m R_\oplus^5 k}{2r^4(f)r^3(f-\tau)}(3\cos^2\tau + D)$$

$$s = \frac{3}{2}\frac{\gamma m R_\oplus^5 k}{r^4(f)r^3(f-\tau)}\frac{\pm q'\sin 2\tau}{\sqrt{1-q^2\sin^2\phi}}.$$

Here, γ is the gravitational constant; $R_\oplus$ is the radius of the Earth; m is the mass of the Moon; D and k are the constants containing the Love numbers characterizing the inhomogeneity of the internal structure of the Earth and the Moon; τ is the delay of the tidal bulge with respect to the line connecting the centres of the Earth and the Moon; q and q' depend on the angles characterizing the position of the orbital plane; $\phi = \omega + f$, where ω is the angle between the node and the perigee.

This model of the tide provides a good description of the evolution in the system if the attracting bodies are situated far from each other and move along almost circular orbits. When the bodies are closer to each other and have significant eccentricities, another model must be used which takes into account the dependence of the tide on time (or on the phase of the tide - forming body).

2. First we consider a model of the lunar - type tide, when two tidal bulges are formed on the Earth (Fig. 1).

For the sake of simplicity, we disregard the friction and the constant delay component caused by it.

Let M denote the mass of the bigger body, m the mass of the smaller body, and m_1 the mass of the bulge; $R_\oplus$ is the radius of the bigger body, r_0 is the distance between the centres; r_1 and r_2 are the distances from the centre of the tide-forming body to the corresponding tidal bulge; α is the angle between the straight line connecting centres of the bodies and the segment r_1 (or r_2).

If the tide - forming body moves along its orbit with a significant eccentricity, then the tidal bulges move in the variable gravitation field

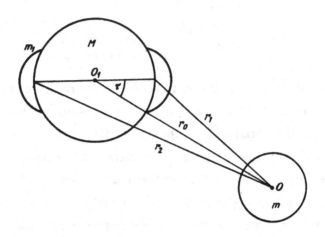

Figure 7.1: Lunar – type tide.

which can bring about a parametric excitation of the oscillations of the tidal bulge phase.

The projection F_x of the attracting force of the tidal bulge onto the normal to the line connecting the bodies can be represented as

$$F_x = -\gamma m m_1 \left(\frac{1}{r_1^2} - \frac{1}{r_2^2} \right) \sin \alpha =$$

$$= \gamma m m_1 \frac{4 R_\oplus^2 \sin \tau \cos \tau}{r_0^4} = \gamma m m_1 \frac{2 R_\oplus^2}{r_0^4} \sin 2\tau.$$

For small oscillations ($|\tau| \ll 1$) we obtain the linearized equation

$$\ddot{\tau} + 2 \frac{\gamma m}{r_0^4} R_\oplus \tau = 0. \qquad (4.2)$$

We substitute into (4.2) the expression for the known elliptical motion

$$r_0 = \frac{p}{1 + e \cos f}, \qquad (4.3)$$

where e is the eccentricity of the orbit of the body m, and p is the orbital parameter ($\sim r_0$). We expand r_0^{-4} in Eq. (4.2) into powers of the eccentricities

$$\ddot{\tau} + \frac{2\gamma m R_{\oplus}}{p^4}(1 - 4e\cos f)\tau = 0. \tag{4.4}$$

The latter equation describes the dynamics of the delay variable τ.

3. It is interesting to compare the periods of the oscillations and rotations

$$T_{osc} = \frac{2\pi}{\sqrt{\gamma m}}\frac{p^2}{\sqrt{R}}, \quad T_{rot} = \frac{2\pi}{\sqrt{\gamma M}}p^{\frac{3}{2}}m,$$

$$d = \frac{T_{osc}}{T_{rot}} = \sqrt{\frac{M}{m}}\sqrt{\frac{p}{R}}. \tag{4.5}$$

For the Earth-Moon system $\delta \approx 50$, which means that the parametric excitation in Eq. (4.4) becomes weak, since it occurs on higher order resonance. In the Gauss evolutionary equation the resonance is possible for high-order harmonics only; i.e. these processes have only an insignificant effect on the evolution of the orbital parameters.

The ratio d will be substantially smaller for more compact systems such as double stars [82], since the ratio of the star masses is of the order one. For double stars, the ratio r_0/R is typically also of the order one. According to (4.5), the ratio of the periods of oscillations and rotations depends on the square roots of their values. This also brings oscillation and rotation frequencies closer together.

Therefore, for such systems, the parametric excitation in Eq. (4.4) becomes effective, and the resonances in the Gauss evolutionary equation (4.1) are observed on the first harmonics, which brings about evolutionary effect for semi-axe and other orbital parameters.

4. For more compact systems, however, more typical is yet another model of the tide in which there is only one pear-shaped tidal bulge (Fig. 2).

Retaining the previous notations, we write a linearized equation for τ, disregarding the friction, which, generally speaking, change the frequency of the process:

$$\ddot{\tau} + \frac{\gamma m}{p^3}(1 - 3e\cos f)\tau = 0. \tag{4.6}$$

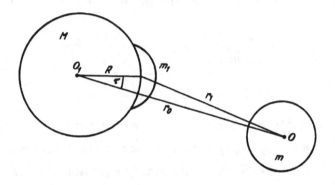

Figure 7.2: Tide in compact system.

For the model model under consideration

$$d = \sqrt{\frac{M}{m}}.$$

Let us now introduce a new independent variable f in Eqs. (4.1) and (4.6). Since f and t are related by the equation (see [1])

$$\frac{\mathrm{d}f}{\mathrm{d}t} = \frac{\sqrt{Gp}}{p^2}(1 + e \cos f)^2,$$

with $G = \gamma(M + m)$, we have with accuracy e^2

$$\frac{\mathrm{d}a}{\mathrm{d}f} = \frac{A R_\oplus^5 p^2}{r_0^4(f) r_0^3(f - \tau)}[(3 \cos 2\tau + D')e \sin f \pm$$

$$\pm \frac{1 - 2e \cos f}{\sqrt{1 - q^2 \sin^2 \phi}} q' \sin 2\tau,$$

where $A = const$, $D' = const$, the delay τ is variable and is determined from the equation

$$\frac{\mathrm{d}^2\tau}{\mathrm{d}f^2} - 2e \sin f \frac{\mathrm{d}\tau}{\mathrm{d}f} + \frac{m}{M + m}(1 - 7e \cos f)\tau = 0. \qquad (4.7)$$

Let us introduce the small parameter

$$\mu = \frac{R_{\oplus}^5}{p^5}.$$

Thus, for the slow variable a we get the equation in the standard form with a deviating argument (3.1):

$$\frac{da}{df} = \mu \frac{Ap^7}{r_0^4 r_0^3(f - \tau)} [(3cos2\tau + D)e \sin f \pm$$

$$\pm \frac{q'(1 - 2e \cos f) \sin 2\tau}{\sqrt{1 - q^2 \sin^2 \phi}} \Bigg],$$

where the delay variable $\tau(f)$ is determined from Eq. (4.7).

Let us substitute the expression (4.3) into r_0. Then,

$$\frac{da}{df} = \mu A [(3cos2\tau + D)e \sin f \pm$$

$$\pm \frac{q'(1 + 2e \cos f) \sin 2\tau}{\sqrt{1 - q^2 \sin^2 \phi}} \pm \frac{q'3e \cos(f - \tau) \sin 2\tau}{\sqrt{1 - q^2 \sin^2 \phi}} \Bigg], \qquad (4.8)$$

$$\ddot{\tau} + \omega_0^2 \tau = e[2\dot{\tau} \sin f + 7\tau \cos f], \qquad (4.9)$$

where

$$\omega_0 = \frac{m}{M + m}.$$

The equation (4.9) describes the dynamics of the tide delay in the model, with friction not taken into account.

As can be seen from Eqs. (4.11) and (4.9), the effective parametric excitation of the tidal bulge oscillations for not very large d is possible for smaller resonances. The oscillatory process of delay $\tau(f)$ and, therefore, the oscillation of the functions $\sin 2\tau$ and $\cos 2\tau$ causes resonances on the RHS of (4.8) also for smaller harmonics and, in this connection, leads to the evolutionary change of the semi-axis a and other system parameters described by Gauss equation.

CHAPTER 8

Stability of Partial Differential Equations

In this chapter, the generalization of the second Lyapunov method proposed in Chapter 2, is extended to linear partial differential equations containing small nonlinearities. Theorems on stability are proved, and applied problems are considered.

8.1 Statement of the Problem

We consider stability of linear equations with small perturbations which are, generally speaking, nonlinear [52] . As to the unperturbed equation, we assume that its zero solution is stable only, i.e. the asymptotic stability of the zero solution does not hold. This means that the generating (unperturbed) operator belongs to a neutral type, when its fundamental operator is uniformly bounded in the operator norm for any t. If the generating operator is independent of time, the intersection of its spectrum with the imaginary axis is permitted. The ideas of the generalized Lyapunov method can be extended to these classes of equations. The stability properties of the generating equation are described by means of positive – definite functionals similar to the Lyapunov functions for ordinary differential equations. Perturbed functionals satisfying general conditions will be considered for equations with perturbations. Restriction on the sign of derivative is understood in terms of the mean calculated by integration along the solutions of the generating equation. The technique described below may be used to investigate stability of a rather broad class of perturbed linear equations which are described by closed operators.

As an example of applied problems, we consider a system of nonlinear acoustic equations.

Let us consider the Cauchy problem

$$\dot{x} = A(t)x + \mu F(t,x), \quad x(t_0) = x_0. \tag{1.1}$$

The unperturbed equation (with the same initial data) has the form

$$\dot{x} = A(t)x, \quad x(t_0) = x_0, \tag{1.2}$$

where $A(t)$ is a closed operator with the domain $D(A) \subset B$ independent of t and dense in the Banach space B, i.e. $\overline{D(A)} = B$.

The equations (1.1) and (1.2) are considered in the region

$$P = \{x : \| x \|_B < N, \ t \geq 0\}. \tag{1.3}$$

The perturbation $\mu F(t,x)$ is such that $\forall t \in J$, $J = [0, \infty)$, $F(t,x)$ maps the Banach space into itself, i.e. $B \rightarrow B$; $x(t)$ is an abstract function in the space B. We assume that for $\phi \in [0, \mu_0]$, the Cauchy problem ($x_0 \in D(A)$) for Eqs. (1.1) and (1.2) is correctly posed. Then, for any $D(A)$ the Cauchy problem has a unique solution. We also assume that the solution of Eqs. (1.1) and (1.2) can be extended to the semi-axis $J = [0, \infty)$.

We introduce the notation $U(t, \tau)$, which is a resolving operator of the unperturbed equation (1.2) and $x = \bar{x}(t, x_0, t_0)$ is the solution of Eq. (1.2).

Lemma 8.1 *Let us assume that the Cauchy problem for Eqs.* (1.1) *and* (1.2) *is correctly posed on* $D(A)$, *where* $\overline{D(A)} = \mathcal{B}$. *Suppose that in the region* P *there exists a function* $L(t, \tau)$ *such that*

$$\| U(t, \tau)F(t, \tau) \| \leq L(t, \tau),$$

where

$$\int_{t_1}^{t_2} L(t, \tau)d\tau \leq L_0(t_1 - t_2)$$

for any finite segment $[t_1, t_2]$.

Then, for any finite l *and solutions* $x = x(t)$, $x = \bar{x}(t)$ *of Eq.* (1.1) *and* (1.2), *with thwe same initial data* $x(t_0) = \bar{x}(t_0) = x_0$ *the estimate*

$$\| x(t) - \bar{x}(t) \| \leq \mu L_0 l$$

is valid for $t \in [t_0, t_0 + l]$.

Proof. According to the definition of the resolvent operator $U(t, \tau)$, we have

$$\bar{x}(t) = U(t, t_0)x_0,$$
$$x(t) = U(t, t_0)x_0 + \mu \int_{t_0}^{t} U(t, \tau)F(\tau, x(\tau)) \, d\tau.$$

Therefore,

$$x(t) - \bar{x}(t) = \mu \int_{t_0}^{t} U(t, \tau)F(\tau, x(\tau)) \, d\tau.$$

From here, and from the condition of the lemma for $t \in [t_0, t_0 + l]$ it follows that

$$\| x(t) - \bar{x}(t) \| \leq \mu \int_{t_0}^{t} \| U(t, \tau)F(\tau, x(\tau)) \| \, d\tau. \leq$$
$$\leq \mu \int_{t_0}^{t} L(t, \tau) \, d\tau \leq \mu L_0 l.$$

The lemma is proved.

In what follows, we assume that $\mathcal{B}$ is a Hilbert space ($\mathcal{B} = \mathcal{H}$); here, as a rule, $\mathcal{H}$ is a functional space of n-dimensional functions ($\mathcal{H} =$

$\mathcal{H}(L^n)$) and, as a Hilbert space, it is complete in the norm generated by a scalar product.

We introduce the following notations:

1. $\mathcal{L}(\mathcal{H}, \mathcal{H})$ is a Banach space of bounded operators acting from the space $\mathcal{H}$ into $\mathcal{H}$.

2. $\Lambda_t(\mathcal{H})$ is a set of bounded positive-definite operator – valued functions $V(t) \in \mathcal{L}(\mathcal{H}, \mathcal{H})$, $\forall t$ acting in the space and satisfying the conditions:

  a) $\| V(t) \| < c$, $(c = const)$; $V = V^*$;

  b) $\exists V^{-1}(t)$ and $\| V^{-1}(t) \| \leq c$;

  c) $V(t)$ is a Bochner differentiable operator – valued function;

  d) for any $\delta > 0$ there exists $\gamma(\delta) > 0$ such that $|(V(t)x, x)_{\mathcal{H}}| < \delta$ when $\| x \| < \gamma$; in particular, there exists a constant α_1 such that $\alpha_1(x, x)|_{\mathcal{H}} \leq (V(t)x, x)_{\mathcal{H}}$;

  e) $V D(A) \subset D(A)$.

3. $S_R^H(z)$ is a ball of the radius R with the centre at the point z; $\bar{S}_R^H(z)$ is a closed ball.

4. $R(l_1, l_2) = \{x : l_1 < \| x \| < l_2\}$ is an annular region; $\bar{R}(l_1, l_2) = \{x : l_1 \leq \| x \| \leq l_2\}$ is a closed annular region.

5. $f(x)$ is a functional on elements of the space H; $\Omega_f\{c_0\} \equiv \{x \in h : f(x) = c_0\}$ is a manifold; $\Omega_f^t\{c_0\} \equiv \{x \in h : f(x, t) = c_0\}$ is a moving manifold;

6.

$\text{Int}(\Omega_f\{c_0\} \equiv \Omega_f\{c_0\} \cup \text{int}(\Omega_f\{c_0\}) = \{x : f \leq c_0\}$;

$\text{Int}(\Omega_f^t\{c_0\} \equiv \Omega_f^t\{c_0\} \cup \text{int}(\Omega_f^t\{c_0\}) = \{x : f(t, x) \leq c_0\}$;

7. $v_0(t, x) = (V(t)x, x)_{\mathcal{H}}$ is a functional $V(t) \in \Delta_t(\mathcal{H})$.

8. $v = v_0(t, x) + u(t, x, \mu, \epsilon) = (V(t)x, x)_{\mathcal{H}} + u(t, x, \mu, \epsilon)$ s a perturbed functional $V(t) \in \Delta_t(\mathcal{H})$. The functional $u(t, x, \mu, \epsilon)$ for small μ can be chosen sufficiently small. We assume that this functional s definite and differentiable with respect to x and t in the annulus $\bar{R}(l_1, l_2)$; ϵ is the size of the neighbourhood whose stability is to be

investigated.

If in Eq. (1.1)

$$F(t, 0) \equiv 0,$$

then this equation has the trivial solution.

Definition 8.1 *The trivial solution of Eq.* (1.1) *is said to be Lyapunov-stable if for any* $\epsilon > 0$ *there exists* $\delta(\epsilon) > 0$ *such that, if* $x_0 \in S_\delta \cap D(A)$, *then* $\parallel x(t, t_0, x_0, \mu) \parallel < \epsilon$ *for* $t \geq t_0$.

The trivial solution is called *unstable* if it is not stable.

Definition 8.2 *The trivial solution of Eq.* (1.1) *is called* (x, μ)-*stable if for any* $\epsilon > 0$ *there exist* $\eta(\epsilon)$ *and* $\mu_0(\epsilon)$ *such that, if* $x_0 \in S_\delta \cap D(A)$, $0 < \mu < \mu_0(\epsilon)$, *then* $\parallel x(t, t_0, x_0, \mu) \parallel < \epsilon$ *for* $t \geq t_0$.

We set $\bar{\epsilon} > 0$ ($\bar{\epsilon} < \epsilon_0$). We select $w_0 > 0$ such that $\Omega_{\alpha_1(x,x)_\mathcal{H}}\{w_0\} \equiv \bar{S}_{(w_0/\alpha_1)^{\frac{1}{2}}}(0) \subset S_{\bar{\epsilon}}(0)$. We set $\sigma > 0$, with the condition $2\sigma < w_0$. Then, since $V(t) \in \Delta_t(H)$ and since the functional $V(t)$ is positive - definite, we have

$$\Omega^t_{v_0}\{w_0\} \subset \text{Int}(\Omega_{\alpha_1(x,x)_\mathcal{H}}\{w_0\}),$$
$$\Omega^t_{v_0}\{w_0 - 2\sigma\} \subset \text{Int}(\Omega_{\alpha_1(x,x)_\mathcal{H}}\{w_0\}). \qquad (1.4)$$

By virtue of the properties of the functional $V(t)$, $v_0 = (V(t)x, x)_\mathcal{H}$ is a positive - definite functional such that

$$\forall \delta > 0, \ \exists \gamma > 0 : \ \parallel x \parallel < \gamma \Rightarrow |v_0| < \delta.$$

Hence, it follows that there exists $\eta(\epsilon) > 0$ such that

$$S_\eta \subset \text{Int}(\Omega^t_{v_0}\{w_0\}). \qquad (1.5)$$

This property of the functional w_0 is similar to the property of the Lyapunov function, i.e. it has an infinitely small upper bound.

Lemma 8.2 *Let the perturbation $u(t, x, \mu, \epsilon)$ be continuous and let its modulus be bounded by the number $\frac{1}{2}\sigma$ (where σ satisfies $2\sigma < w_0$) in the annulus $\bar{R}(\eta, \bar{\epsilon})$, $t > 0$. Then $\Omega^t_{v_0+u}\{w_0 - \sigma\} \subset \bar{R}(\eta, \bar{\epsilon})$ and the manifold $\Omega^t_{v_0+u}\{w_0 - \sigma\}$ is closed.*

Proof. In the annulus $\bar{R}(\eta, \bar{\epsilon})$, the positive - definite functional $v_0(t, x) = (V(t)x, x)_{\mathcal{H}}$ takes on the manifold $\Omega^t_{v_0}\{w_0 - 2\sigma\}$ the value $w_0 - 2\sigma$, while on the manifold $\Omega^t_{v_0}\{w_0\} \subset \bar{R}(\eta, \bar{\epsilon})$ it takes the value w_0. Thus, in $\bar{R}(\eta, \bar{\epsilon})$ we have

$$(V(t)x, x)_{\mathcal{H}} = w_0 - 2\sigma \quad \text{on the manifold} \quad \Omega^t_{v_0}\{w_0 - 2\sigma\}, \qquad (1.6)$$

$$(V(t)x, x)_{\mathcal{H}} = w_0 \quad \text{on the manifold} \quad \Omega^t_{v_0}\{w_0\}. \qquad (1.7)$$

The continuous functional $v = v_0 + u$ on the manifold $\Omega^t_{v_0}\{w_0 - 2\sigma\}$ takes the value v^* which satisfies the inequalities

$$w_0 - 2\sigma - \frac{1}{2}\sigma \le v^* \le w_0 - 2\sigma + \frac{1}{2}\sigma. \qquad (1.8)$$

The value of the functional v on the manifold $\Omega^t_{v_0}\{w_0\}$ will be denoted v^{**},

$$w_0 - \frac{1}{2}\sigma \le v^{**} \le w_0 + \frac{1}{2}\sigma. \qquad (1.9)$$

Since $v = v_0 + u$ is continuous in the region $\bar{R}(\eta, \bar{\epsilon})$ it takes an intermediate value $w_0 - \sigma$. Thus, $\Omega^t_{v_0+u}\{w_0 - \sigma\} \subset \bar{R}(\eta, \bar{\epsilon})$. The first assertion of the lemma is proved.

Let us construct an arbitrary Jordan curve $C(t)$ from a point of the manifold (sphere)

$$\Omega_{(x,x)_{\mathcal{H}}}\{w_0\} \equiv \{x : \| x \| = \eta\}$$

to the point of the manifold

$$\Omega_{(x,x)_{\mathcal{H}}}\{w_0\} \equiv \{x : \| x \| = (w_0/\alpha_1)^{\frac{1}{2}}\}.$$

On this Jordan curve there is a point C^* such that

$$C^* \in \Omega^t_{v_0}\{w_0 - 2\sigma\}, \quad v|_{C^*} = v^*,$$

where v^* satisfies the inequality (1.8). On the same path there is a point C^{**} such that

$$C^{**} \in \Omega_{v_0}^t\{w_0\}, \quad v|_{C^{**}} = v^{**},$$

where v^{**} obeys the inequality (1.9). Since the curve $C(t)$ on the segment $C(t')$ between C^* and C^{**} is continuous, there exists a point $\tilde{C}$ such that $v|_{\tilde{C}} = w_0 - \sigma$, as $v^* < w_0 - \sigma < v^{**}$.

Since $C(t')$ is arbitrary, the equation

$$(V(t)x, x)_{\mathcal{H}} + u(t, x, \mu, \epsilon) = w_0 - \sigma$$

defines a closed manifold $\Omega_{v_0+u}^t\{w_0 - \sigma\}$, with

$$\Omega_{v_0+u}^t\{w_0 - \sigma\} \subset \bar{R}(\nu, \bar{\epsilon}),$$
$$\Omega_{v_0+u}^t\{w_0 - \sigma\} \subset \Omega_{\alpha_1(x,x)_{\mathcal{H}}}^t\{w_0\} \equiv \bar{S}(w_0/\alpha_1)^{\frac{1}{2}}(0).$$

The lemma is proved.

8.2   Theorem on Stability

Let $A(t)$ be a closed operator with the domain $D(A)$ dense in $\mathcal{H}$, $\overline{D(A)} = \mathcal{H}(D(A) \subset D(A^*))$. Let us now calculate the derivative of the functional $v = v_0 + u$ by virtue of Eq. (1.1):

$$\dot{v} = (\Phi x, x)_{\mathcal{H}} + \theta(\mu)\phi(t, x, \mu, \epsilon), \qquad (2.1)$$

where

$$\Phi = \dot{V} + VA + A^*V,$$
$$\theta(\mu)\phi(t, x, \mu, \epsilon) =$$
$$2\mu \mathrm{Re}(Vx, F)_{\mathcal{H}} + \frac{\partial u}{\partial t}\left(\frac{\partial u}{\partial x}, Ax\right)_{\mathcal{H}} + \mu\left(\frac{\partial u}{\partial x} - F\right)_{\mathcal{H}} =$$
$$= \mu(F, Vx)_{\mathcal{H}} + \mu(Vx, F)_{\mathcal{H}} + \frac{\partial u}{\partial t} + \left(\frac{\partial u}{\partial x}, Ax\right)_{\mathcal{H}} + \mu\left(\frac{\partial u}{\partial x}, F\right)_{\mathcal{H}}.$$

The factor $\theta(\mu)$ is singled out to ensure that the functional $\phi(t, x, 0, \, e)$ does not become identically zero.

The functional $(\Phi(t)x, x)_{\mathcal{H}}$ is the derivative of the functional $v_0 = (V(t)x, x)_{\mathcal{H}}$, $V(t) \in \Delta_t(\mathcal{H})$. By virtue of Eq. (1.2), the function $x(t)$: $\forall t \in J$, $x(t) \in D(A)$ and $V(t)D(A) \subset D(A)$ being a solution of Eq. (1.1) or (1.2).

To formulate the main theorem on stability, we make the following assumptions:

a) the space $\mathcal{B}$ which contains the domain of the operator $A(t)$, is a Hilbert space $(\mathcal{B} = \mathcal{H})$;

b) the Cauchy problem for Eqs. (1.1) and (1.2) is correctly posed on $D(A)$; in Eqs. (1.1) and (1.2) the operator – valued function $A(t)$ is closed and has a domain $D(A) = \mathcal{H}(D(A) \subset D(A^*))$ independent of t;

c) there exists an operator $V(t) \in \Delta_t(\mathcal{H})_\beta$ such that

c') $\forall \beta > 0$, $\exists \gamma > 0$; $\parallel x \parallel < \gamma$, $t > 0 \Rightarrow |V_0| < \beta$;

c") the operator $\Phi(t) = \dot{V} + VA + A^*V$ is dissipative in $D(A) \subset \mathcal{H}$, i.e. $(\Phi x, x)_{\mathcal{H}} \le 0$ $\forall x \in D(A)$, $D(A) = \mathcal{H}$ or, in the general case, $(\dot{V}x, x) + (VAx, x) + (V, Ax) \le 0$ for all $x \in D(A)$;

d) for any $\epsilon > 0$ $(\epsilon < \epsilon_0 < N)$, there exist w, $\sigma_0 < w_0/2$, $\gamma = \gamma(w_0, \sigma_0)$ (related as indicated in Section 1) such that $\forall \rho \ge \gamma$ $(\rho < \epsilon)$ in the annular region $R(\rho, \epsilon)$, $t > 0$ the functionals $u(t, x, \mu, \epsilon)$, $\phi(t, x, \mu, \epsilon)$ are defined and exist;

d') $\delta > 0$ and $l > 0$ such that if for $t \in [t_0, t_0 + T]$ we have $\bar{x}(t) \in R(\rho, \epsilon)$ (where $\bar{x}(t)$ is a solution of Eq. (1.2) and $T > l$), then, uniformly with repect to $t_0 > 0$

$$\int_{t_0}^{t_0+T} \phi(t, \bar{x}(t), \mu, \epsilon)\, dt < -\delta T;$$

d") there exist summable functions $F(t)$, and $L(t, \tau)$, constants F_0, L_0 and a non-decreasing function $\chi_1(\alpha)$, $\lim_{\alpha \to 0} \chi_1(\alpha) = 0$, such that for > 0 and $x \in R(\rho, \epsilon)$

$$\parallel \phi(x', t) - \phi(x'', t) \parallel \le \chi_1(\parallel x' - x'' \parallel)F(t),$$

$$\| U(t,\tau)F(t,\bar{x}(t)) \| \le L(t,\tau),$$

$$\int_{t_1}^{t_2} F(t)\, dt \le F_0(t_2 - t_1),$$

$$\int_{t_1}^{t_2} L(t,\tau)\, d\tau \le L_0(t_2 - t_1)$$

on any finite segment $[t_1, t_2]$;

d''') *there exists a non-decreasing function* $\chi_2(\alpha)$, $\lim_{\alpha \to 0}\chi_2(\alpha) = 0$, *such that*

$$|u(t, x, \mu, \epsilon)| < \chi_2(\mu).$$

Theorem 8.1 *Suppose that assumptions (a) to (d) are satisfied.*

Then there exist $\eta(\epsilon)$ *and* $\mu(\epsilon)$ *such that any solution of Eq.* (1.1) *with the initial data* $x(0) = x_0$, $x_0 \in S_\eta(0) \cap D(A)$ *for* $\mu < \mu(\epsilon)$ *satisfies the inequality* $\| x(t) \| < \epsilon$ *for all* $t > 0$.

Proof. Let us set $\bar{\epsilon} > 0$ ($\bar{\epsilon} < \epsilon$) and consider the manifolds $\Omega_{v_0}^t\{w_0\}$ and $\Omega_{v_0}^t\{w_0 - 2\sigma\}$. If $\sigma > 0$ ($\sigma < \sigma_0$) is chosen, then, according to condition (c') of the theorem, there exists $\eta \ge \gamma$ such that $S_\eta(0) \subset \mathrm{Int}(\Omega_{v_0}^t\{w_0\})$, $S_\eta(0) \subset \mathrm{Int}(\Omega_{v_0}^t\{w_0 - 2\sigma\})$.

In this case assumptions d'-d''' will be satisfied in $R(\eta, \epsilon)$. We consider the manifold $\Omega_{v_0+u}^t\{w_0-2\sigma\}$. Condition (d''') makes it possible to choose μ_1 so small that for all μ, $0 < \mu < \mu_1$, the inequality

$$|U(t, x, \mu, \epsilon)| < \frac{1}{2}\sigma$$

holds.

Then, according to Lemma 8.2, this manifold is closed, and

$$\Omega_{v_0+u}^t\{w_0 - 2\sigma\} \subset R(\eta, \epsilon), \quad t > 0.$$

Let $x = x(t)$ be a solution of Eq. (1.1) such that

$$x(0) \in S_\eta(0) \cap D(A).$$

Suppose that this solution left the sphere $\bar{S}_\eta(0)$ and

$$t_0 = \min\{t : x(t) \in \Omega^t_{v_0+u}\{w_0 - \sigma\}\}$$

$$(v_0 = (V(t)x, x))$$

We differentiate $v = v_0 + u$ along this solution $x = x(t)$. The derivative will be expressed by (2.1). The operator $\Phi(t)$ is dissipative by condition (c"); this means that the trivial solution of the unperturbed equation (1.2) is Lyapunov-stable (Definition 8.1). Indeed, since the operator $\Phi(t) = \dot{V} + VA + A^*V$ is dissipative, i.e.

$$\dot{v}_0 = (\Phi(t)\bar{x}, \bar{x})_\mathcal{H} \leq 0, \quad \bar{x} \in D(A),$$

we have

$$\alpha_1 \parallel \bar{x}(t, x_0) \parallel^2 \leq (V(t)\bar{x}(t, x_0), \bar{x}(t, x_0)_\mathcal{H} \leq (V(0)x_0, x_0)_\mathcal{H}.$$

According to the properties of the operator $V(t)$ $(V(t) \in \Delta_t(\mathcal{H}))$, if the norm $\parallel x_0 \parallel$, $(x_0 \in D(A))$ is sufficiently small, the expression $(V(0)x_0, x_0)_\mathcal{H}$ can be made arbitrarily small as well, hence, the point $x = 0$ of Eq. (1.2) is stable.

Thus, having integrated Eq. (2.1) along the solution $x = x(t)$ and taking into account the dissipativity of $\Phi(t)$ on $D(A)$, we obtain

$$v(t, x(t)) \leq v(t_0, x_0) + \theta(\mu) \int_{t_0}^t \phi(\tau, x(\tau), \mu, \epsilon) \, d\tau. \qquad (2.2)$$

We construct the solution $\bar{x}(t)$ of the unperturbed equation (1.2) which starts from the same point t_0, $x_0 \in J \otimes D(A)$. Using this solution, we represent the integral in (2.2) as follows:

$$\int_{t_0}^t \phi(\tau, x(\tau), \mu, \epsilon) \, d\tau = \int_{t_0}^t \phi(\tau, \bar{x}(\tau), \mu, \epsilon) \, d\tau +$$

$$\qquad (2.3)$$

$$+ \int_{t_0}^t [\phi(\tau, x(\tau), \mu, \epsilon) - \phi(\tau, \bar{x}(\tau), \mu, \epsilon)] \, d\tau.$$

Thus, according to assumption (d') of the theorem, there exist $\delta > 0$ and $l > 0$ such that, if $t > t_0 + l$ and if for $\tau \in [t_0, t]$ the solution $x = \bar{x}(\tau)$ is located in the annular region $\bar{x}(\tau) \in R(\eta, \epsilon)$, then uniformly with respect to t_0

$$\int_{t_0}^{t} \phi(\tau, \bar{x}(\tau), \mu, \epsilon) \, d\tau \leq -\delta(t - t_0). \tag{2.4}$$

Let us now examine variation of v if $t \in [t_0, t_0 + l]$, where l is defined by condition (d') of the theorem. Condition (c) ensures stability of the trivial solution of the unperturbed equation (1.2), thus, the solution $\bar{x}(t)$ for $t > t_0$ will not leave the ϵ-neighbourhood of zero.

If we choose the small parameter μ, Lemma 8.1 guarantees that the difference between the solutions of the perturbed and unperturbed equations is smaller than $\epsilon - \bar{\epsilon}$ on a segment of the length l. It follows that the particular solution $x = x(t)$ of Eq. (1.1) does not leave the ϵ-neighbourhood $S_\epsilon(0)$ of the trivial solution $x = 0$. Let

$$\mu_2 = \frac{\epsilon - \bar{\epsilon}}{L_0 l}.$$

According to Lemma 8.1, for $0 < \mu < \mu_2$, $t \in [t_0, t_0 + l]$

$$\| x(t) - \bar{x}(t) \| \leq \epsilon - \bar{\epsilon}.$$

Then either the solution $\bar{x}(t)$ remains outside the η-neighbourhood $S_\eta(0)$ for $t_0 \leq t \leq t_0 + l$, or there exists a moment t_1, $t_1 \in [t_0, t_0 + l]$ such that $\| \bar{x}(t) \| \leq \eta$.

We begin with the second possibility. In this case we have the inequality

$$(V(t_1)\bar{x}(t_1), \bar{x}(t_1))_{\mathcal{H}} < w_0 - 2\sigma$$

and for the functional v we have

$$\begin{aligned}
v(t_1, x(t_1)) &= (V(t_1)x(t_1), \bar{x}(t_1))_{\mathcal{H}} + u = \\
&= (V(t_1)\bar{x}(t_1), \bar{x}(t_1))_{\mathcal{H}} + u + (V(t_1)x(t_1), x(t_1))_{\mathcal{H}} - \\
&\quad - (V(t_1)\bar{x}(t_1), \bar{x}(t_1))_{\mathcal{H}}, \quad \left(|u| < \frac{1}{2}\sigma \right).
\end{aligned}$$

To estimate the latter difference, we use continuity of the functional $v_0 = (Vx, x)_{\mathcal{H}}$ and Lemma 8.1. We choose $\bar{\mu}_2$ small enough to ensure the inequality

$$|(V(t_1)x(t_1), x(t_1))_{\mathcal{H}} - (V(t_1)\bar{x}(t_1), \bar{x}(t_1))_{\mathcal{H}}| < \frac{1}{2}\sigma$$

for $\mu < \bar{\mu}_2$ and $t_1 \in [t_0, t_0 + l]$. Then $v(t_1, x(t_1)) < w_0 - \sigma$ and the solution $x = x(t)$ for $t = t_1$ is inside the initial manifold,

$$x(t_1) \in \Omega_{v_0}^t\{w_0 - \sigma\}.$$

We shall consider the case in which the solution remains outside the η-neighbourhood of the point $x = 0$.

According to condition (d") of the theorem,

$$|\phi(t, x(t)) - \phi(t, \bar{x}(t))| \le \chi_1(\| x - \bar{x} \|)F(t).$$

Applying Lemma 8.1, we obtain

$$\int_{t_0}^t |\phi(\tau, x(\tau)) - \phi(\tau, \bar{x}(\tau))| \le \chi_1(\mu L_0 l)F_0(t - t_0).$$

We choose μ_3 so small that the inequality

$$\chi_1(\mu L_0 l)F_0 \le \frac{1}{2}\delta$$

s satisfied. Then, on the segment $[t_0, t_0 + l]$ we have

$$\int_{t_0}^t |\phi(\tau, x(\tau)) - \phi(\tau, \bar{x}(\tau))| \le \frac{1}{2}\delta(t - t_0). \qquad (2.5)$$

Let $\mu_0 = \min\{\mu_1, \mu_2, \bar{\mu}_2, \mu_3\}$. Then for $0 < \mu < \mu_0$ and $t \in [t_0, t_0 + l]$ he inequalities $\| x(t) \| < \epsilon$, $|u| < \frac{1}{2}\sigma$ and (2.5) hold. Combining (2.5), condition (d') and relations (2.3), (2.4), we estimate $\int_{t_0}^t \phi(\tau, x(\tau))\delta\tau$ for $\in [t_0 + l, \infty)$:

$$\int_{t_0}^t \phi(\tau, x(\tau))\delta\tau \le \left(-\delta + \frac{1}{2}\delta\right)(t - t_0) = -\frac{1}{2}\delta(t - t_0).$$

Thus, the solution $x = x(t)$ remains in the ϵ-neighbourhood $S_\epsilon(0)$ for $t \in [t_0, t_0 + l]$. Along this solution, starting from a point on the segment $[t_0, t_0 + l]$, the integral $\int_{t_0}^{t} \phi(\tau, x(\tau))\delta\tau$ is negative, thus, the functional v decreases. This means that the solution returns into the manifold

$$\Omega_v^t\{w_0 - \sigma\} = \{x \in \mathcal{H} : (V(t)x, x)_{\mathcal{H}} + u(t, x, \mu, \epsilon) = w_0 - \sigma\}.$$

All these estimates are uniform with respect to t_0, $x_0 \in D(A)$ thus, $\| x \| \leq \epsilon$ for all $t > 0$. The theorem is proved.

Remark. If the operator $\Phi(t)$ is uniformly dissipative, i.e. $(\Phi\bar{x}, \bar{x})_{\mathcal{H}} \leq \omega(\bar{x}, \bar{x})_{\mathcal{H}}$, where $\omega > 0$ is a constant, then for sufficiently small norm of x_0 the solution $\bar{x}(t, x_0)$ tends to zero in the norm, as $t \to +\infty$, i.e zero attracts; thus the trivial solution of Eq. (1.2) is asymptotically Lyapunov-stable.

If $u \equiv 0$, we have a particular case of Theorem 8.1.

Theorem 8.2 . *Suppose that assumptions (a) to (c) are satisfied. Let the limit*

$$\phi_0(t_0, x_0) = \lim_{T \to \infty} \frac{1}{T} \int_{t_0}^{t_0+T} Re(V(t)F(t, \bar{x}(t)), \bar{x}(t))_{\mathcal{H}} dt$$

exist uniformly with respect to t_0, $x_0 \in D(A)$, and for any $\gamma > 0$ ($\gamma < N$) let there exists $\delta(\gamma) > 0$ such that if $\| x_0 \| > \gamma$, $x_0 \in D(A)$, then $\psi_0(t_0, x_0) < -\delta$ for $t_0 > 0$, where $x = \bar{x}(t)$ is a solution of the unperturbed equation (1.2). Let there also exist summable functions F_0 and $L(t)$, constants F_0 and L_0, and a non-decreasing function $\chi_1(\alpha)$, $\lim_{\alpha \to 0}\chi_1(\alpha) = 0$, such that in the region P we have

$$|Re(V(t)F(t, x'), x')_{\mathcal{H}} - Re(V(t)F(t, x''), x'')_{\mathcal{H}}| \leq$$
$$\leq \chi_1(\| x' - x'' \|)F(t),$$

$$\int_{t_1}^{t_2} F(t)\, dt \leq F_0(t_2 - t_1),$$

$$\| U(t, \tau)F(\tau, x) \| \leq L(t, \tau),$$

$$\int_{t_1}^{t_2} L(t,\tau)\,d\tau \le L_0(t_2 - t_1)$$

on any finite segment $[t_1, t_2]$.

Then, for any $\epsilon > 0$ there exist $\eta(\epsilon)$ and $\mu_0(\epsilon)$ such that if $x(0) = x_0 \in D(A)$ satisfies the inequality $\| x(t) \| < \eta$, then for $\mu < \mu_0(\epsilon)$ and $t > 0$ the inequality

$$\| x(t) \| < \epsilon, \quad x(t) \in S_\epsilon(0).$$

holds.

This theorem follows directly from Theorem 8.1 and has a number of applications.

8.3 Theorem on Stability over Finite Interval

Theorem 8.1 contains the condition that the mean is negative. If this condition is not satisfied, stability over an infinite interval may not occur. In this case the theorem on stability can be proved over a finite interval whose length depends on the value of the small parameter.

Theorem 8.3 *Suppose that assumptions (a), (b), and (c) are satisfied. In addition, for any $\epsilon > 0$ $(\epsilon < \epsilon_0 < N)$, $\rho > 0$ $(\rho < \epsilon)$ in the annulus $R(\rho, \epsilon) = \{x : \rho \le \| x \| < \epsilon\}$, $t > 0$, the functionals $u(t, x, \mu, \epsilon)$, $\phi(t, x, \mu, \epsilon)$ are defined, and exist:*

d') a non-decreasing function $\chi_1(\alpha)$, $\lim_{\alpha \to 0} \chi_1(\alpha) = 0$, such that for $t > 0$ and $x \in R(\rho, \epsilon)$, $x \in D(A)$, we have the estimate

$$|u(t, x, \mu, \epsilon)| < \chi_1(m);$$

d") a constant ϕ_0 such that for $x \in R(\rho, \epsilon)$, $x \in D(A)$ we have the estimate

$$|\phi(t, x, \mu, \epsilon)| < \phi_0.$$

Then there exist, $\sigma(\epsilon)$, $T(\epsilon, \mu) = \sigma(\epsilon)[2\phi_0\theta(\mu)]^{-1}$, $\eta(\epsilon)$, $\mu_0(\epsilon)$ such that all the solutions, for which $x_0 \in D(A) \cap S_\eta(0)$, satisfy the inequality

$$\| x(t) \| < \epsilon$$

for all $0 < t < T(\epsilon, \mu)$, and $\mu < \mu_0(\epsilon)$.

Proof. Let us set $\epsilon > 0$. We choose w_0 such that the 'sphere' $\alpha_1(x, x)_\mathcal{H} = w_0$ lies in $S_\epsilon(0)$.

Let us consider the moving manifolds:

$$\Omega_{v_0}^t\{w_0\} \quad \text{and} \quad \Omega_{v_0}^t\{w_0 - 2\sigma\} \quad (0 < 2\sigma < w_0).$$

Then, by virtue of the property of the functional which is expressed by the inequality

$$\alpha_1(x, x)_\mathcal{H} \leq (V(t)x, x)_\mathcal{H},$$

the manifolds lie inside the sphere

$$\text{Int}(\Omega_{\alpha_1(x,x)_\mathcal{H}}\{w_0\}) \equiv \bar{S}_{(w_0/\alpha_1)^{\frac{1}{2}}}(0).$$

According to condition (c') there exists a neighbourhood $S_\eta(0)$ such that, for any $t > 0$, this sphere lies inside the introduced manifolds.

Let us now consider the manifold

$$\Omega_v^t\{w_0 - \sigma\}$$

and choose μ_0 sufficiently small to ensure the inequality

$$|u(t, x, \mu, \epsilon)| < \frac{1}{4}\sigma$$

for all $\mu < \mu_0$. This can be done by virtue of condition (d'). Let us construct the solution $x = x(t)$ from the region $D(A) \cap S_\eta(0)$ and suppose that at $t = t_0$ this solution intersects the manifold under consideration, i.e. $v|_{t=t_0} = w_0 - \sigma$. According to Lemma 8.2, the manifold

$$\Omega_{v|_{t=t_0}}\{w_0 - \sigma\} \equiv \Omega_v^{t_0}\{w_0 - \sigma\} \tag{3.1}$$

is closed and lies outside the $S_\eta(0)$-neighbourhood of the equilibrium position $x = 0$. Thus the solution $x = x(t)$ can emerge from $S_\epsilon(0)$ only after intersecting the manifold

$$\Omega_v^{t_0}\{w_0 - \sigma\}.$$

Let us consider the derivative of the functional v and use Eq. (1.1) presented in the form (2.1). Integrating Eq. (2.1) from the moment of intersection with the manifold (3.1) and taking condition (c") of Theorem 8.1 into account, we obtain

$$v \le w_0 - \sigma + U,$$

where

$$U(t) = \theta(\mu) \int_{t_0}^{t} \phi(\tau, x(\tau), \mu, \epsilon)\, d\tau.$$

In other words,

$$(V(t)x, x)_\mathcal{H} + u \le w_0 - \sigma + U(t),$$

or

$$v_0 \le w_0 - \sigma - u + U.$$

From the latter inequality and condition (d") of Theorem 8.2 we conclude that for $t \in [t_0, t_0 + T]$ $T = \sigma(\epsilon)[\phi_0\theta(m)]^{-1}$ the inequality

$$|U| \le \frac{1}{2}\sigma$$

holds and, therefore,

$$(Vx, x)_\mathcal{H} \le w.$$

Since the moving manifold $\Omega_{v_0}^{t}\{w_0\}$ lies inside the sphere $\text{Int}(\Omega_{\alpha_1(x,x)_\mathcal{H}}$ $\{w_0\}) = \bar{S}_{(w_0/\alpha_1)^{\frac{1}{2}}}(0)$, the solution $x = x(t)$ for $t \in [t_0, t_0 + T]$ does not leave $S_\epsilon(0)$. The theorem is proved.

8.4   Theorem on Instability

Theorems on instability are an important aspect of applications of perturbed Lyapunov functionals. We consider one of them. For its formulation we need, together with general assumptions (a) and (b) of Theorem 8.1, a number of additional assumptions similar to the further assumptions of Theorem 8.1.

    (c) There exists a functional v_0 such that in any η-neighbourhood of $x = 0$ ($0 < \eta < h$) there exists a subregion where $v_0 > 0$. (Following Chetaev, we denote this subregion by ($v_0 > 0$).) The derivative of the functional v_0 is non - negative by virtue of Eq. (1.2).

    (d) There exist a perturbation $u(t, x, \mu)$ of the functional v_0 and a non-decreasing function $\chi_1(\alpha)$, $\lim_{\alpha \to 0} \chi_1(\alpha) = 0$, such that

$$|u(t, x, \mu)| < \chi_1(\mu)$$

is valid in the region P.

    Let us calculate the derivative of the functional $v = v_0 + u$ and present it in the form (2.1).

    e) There exist summable functions $F(t)$, and $L(t, \tau)$, constants F_0, L_0 and a non-decreasing function $\chi_2(\alpha)$, $\lim_{\alpha \to 0} \chi_2(\alpha) = 0$, such that

$$\| \phi(t, x') - \phi(t, x'') \| \leq \chi_2(\| x' - x'' \|) F(t),$$

$$\| U(t, \tau) F(t, x(t)) \| \leq L(t, \tau),$$

$$\int_{t_1}^{t_2} F(t) \, dt \leq F_0(t_2 - t_1),$$

$$\int_{t_1}^{t_2} L(t, \tau) \, d\tau \leq L_0(t_2 - t_1)$$

in the region $P' \subset P$, $P' = \{(x, t) : \| x \| < h, \ h < N, \ t > 0\}$ where $[t_1, t_2]$ is any finite segment.

    f) The limit

$$\psi(t_0, x_0, \mu) = \lim_{T \to \infty} \frac{1}{T} \int_{t_0}^{t_0 + T} \phi(t, \bar{x}(t)) \, dt$$

exists uniformly with repect to t_0, $x_0 \in P' \cap D(A)$.

The mean $\psi(t_0, x_0, \mu)$ is positive in the region $K(\psi > 0)$ (by the region $K(\psi > 0)$ we understand, as usual, a subregion of the sphere $S_h(0)$ with $\psi > 0$ for $t > 0$).

The intersection of the two regions $(v_0 > 0)$ and $K(\psi > 0)$ will be denoted by $K(v_0, \psi > 0)$.

g) In the region $K(v_0, \psi > 0)$ for any neighbourhood $S_\eta(0)$ $(0 < \eta < h)$, there exist a point x and numbers $k > 0$, $\delta > 0$ such that $\psi(t_0, x_0, \mu) > 0$, $v_0(t_0, x_0) > 2k$ for $t_0 > 0$.

h) There exists $l > 0$ such that on any portion of the solution $x = \bar{x}(t)$, $t \in (t_0, t_0 + l)$ of the unperturbed equation, if this portion belong to P', there exist points from the region $K(v_0, \psi > 0)$ for which $\psi > \delta$, provided that the solution emerges from a point for which $v_0 > k$.

Theorem 8.4 *Suppose that assumptions (a) and (b) of Theorem 8.1, as well as additional assumptions (c) to (h) are fulfilled. Then the trivial solution of Eq. (1.2) is unstable with respect to the perturbation μF.*

Remark 1. Theorem 8.4 includes the important particular case where $u \equiv 0$. In such case, the assumption on the form of the functional $v_0 = (V(t)x, x)_{\mathcal{H}}$ is too retrictive, and it can be relaxed.

Remark 2. The theorems on stability and instability formulated above can be extended in the obvious manner to the case in which stability is investigated not for all but for part of the variables.

Remark 3. Theorems 8.1 to 8.4 determine the conditions of stability with respect to perturbations, when the unperturbed equation (1.2) is linear. These results can be extended to nonlinear equations when the unperturbed equation has a closed operator $f(t, x)$ (depending on as a parameter) such that

$$J \times D(f) \to cb, \quad \bar{D}(f) = \mathcal{B},$$

when the conditions of the existence and uniqueness of a solution of the Cauchy problem $x(0) = x_0 \in D(f)$ are satisfied, and the solution can be extended. In this case, a differentiable functional of the general form can be chosen as v_0. Similar conditions are imposed on the functional and its derivative along the solution of the unperturbed equation.

8.5   Stability of Some Hyperbolic Systems

As an example of the theorems proved above, we consider a hyperbolic system of acoustic equations:

$$\frac{du}{dt} = -\frac{\partial p}{\partial \xi} + \mu f(t)u^{l_1}, \quad l_1 = 2p_1 + 1, \quad p_1 = 0, 1, 2, \dots$$

$$(5.1)$$

$$\frac{dp}{dt} = -\frac{\partial u}{\partial \xi} + \mu g(t)u^{l_2}, \quad l_1 = 2p_2 + 1, \quad p_2 = 0, 1, 2, \dots$$

which contain a small nonlinearity.

The system is defined in the region $0 \le \xi \le 1$, $t \ge 0$ with boundary and initial conditions having the form

$$u(0, t) = u(1, t) = 0,$$
$$u(\xi, 0) = \phi(\xi), \quad p(\xi, 0) = \psi(\xi),$$

where the functions ϕ and ψ satisfy the condition of conjugation with the boundary conditions

$$\phi(0) = \phi(1) = \phi''(0) = \phi''(1) = 0,$$
$$\psi'(0) = \psi'(1) = 0. \tag{5.2}$$

We choose the following functional

$$v_0 = \frac{1}{2}\int_0^1 [u^2(\xi, t) + p^2(\xi, t)]\, d\xi.$$

The unperturbed system admits the solution

$$\bar{u}(\xi, t) = \sum_{k=1}^{\infty} [\imath a_k e^{\imath k\pi t} \sin k\pi\xi - \imath a_{-k} e^{-\imath k\pi t} \sin k\pi\xi],$$

$$\bar{p}(\xi, t) = -\sum_{k=1}^{\infty} [a_k e^{\imath k\pi t} \cos k\pi\xi + a_{-k} e^{-\imath k\pi t} \cos k\pi\xi] - a_0,$$

$$a_s = -\imath \int_0^1 \phi(\xi) \sin s\pi\xi \, d\xi - \imath \int_0^1 \psi(\xi) \cos s\pi\xi \, d\xi,$$

$$s = 0, \pm 1, \pm 2, \ldots$$

The derivative $\dot{v}_0$, calculated by virtue of the unperturbed system, is equal to zero. If the conditions (5.2) are satisfied, then the Cauchy problem for the system (5.1) with $u(\xi, 0) = \phi(\xi)$, $p(\xi, 0) = \psi(\xi)$ ($\mu = 0$) is correctly posed. Having chosen $v = v_0$, i.e. having set the perturbation of the functional v_0 equal to zero, we find the derivative of the functional $v = v_0$ by virtue of the equations of the nonlinear system (5.1) ($\mu \neq 0$):

$$\dot{v}_0 = \frac{1}{2} \int_0^1 [2u\dot{u}_t + 2p\dot{p}_t] \, d\xi =$$

$$\int_0^1 \left[u \left(-\frac{\partial p}{\partial \xi} + \mu f(t) u^{l_1} \right) + p \left(-\frac{\partial u}{\partial \xi} + \mu g(t) p^{l_2} \right) \right] d\xi =$$

$$= -\int_0^1 \frac{\partial}{\partial \xi} (pu) \, d\xi + \mu f(t) \int_0^1 u^{l_1+1} \, d\xi + \mu g(t) \int_0^1 p^{l_2+1} \, d\xi =$$

$$= \mu f(t) \int_0^1 u^{l_1+1} \, d\xi + \mu g(t) \int_0^1 p^{l_2+1} \, d\xi$$

since $u(0, t) = u(1, t) = 0$.

Let us write the mean ψ calculated by integration along the solution of the unperturbed system

$$\psi = \lim_{T \to \infty} \frac{1}{T} \int_{t_0}^{t_0+T} \{ g(t) \int_0^1 [a_0 +$$

$$+ \sum_{k=1}^{\infty} (a_k e^{\imath k\pi t} \cos k\pi\xi + a_{-k} e^{-\imath k\pi t} \cos k\pi\xi)]^{l_2+1} + \tag{5.3}$$

$$f(t) \int_0^1 \sum_{k=1}^{\infty} [\imath a_k e^{\imath k\pi t} \sin k\pi\xi - \imath a_{-k} e^{-\imath k\pi t} \sin k\pi\xi]^{l_1+1} \} \, dt$$

If the function $g(t)$ has, for example, the form $g(t) = g_0 e^{\imath \nu t}$, then it is clear from the expression for the mean ψ (5.3) that if the frequency ν coincides with the frequencies which are multiples of π, then resonances may occur in the system. In the case of resonances the mean ψ may be positive, hence, by applying Theorem 8.4, it can be concluded that the system is unstable.

As a second example we shall consider the system

$$\frac{\partial u}{\partial t} = -\frac{\partial p}{\partial \xi}, \quad \frac{\partial p}{\partial t} = -c_0^2 \frac{\partial u}{\partial \xi} - \mu g(t) \int_0^\xi u^n(t, \eta) \, d\eta, \qquad (5.4)$$

$n = 2l + 1$ ($l > 0$ is an integer) in the region $t \geq 0$, $0 \leq \xi \leq 1$, with the boundary conditions

$$u(0, t) = u(l, t) = 0.$$

This system is related to the wave equation for u:

$$\frac{\partial^2 u}{\partial t^2} - c_0^2 \frac{\partial^2 u}{\partial \xi^2} = \mu g(t) u^n$$

($n > 0$ is odd).

As above, we choose the integral of energy

$$v = v_0 = \frac{1}{2} \int_0^1 \left[u^2(\xi, t) + \frac{p^2(\xi, t)}{c^2} \right] d\xi$$

to define $v = v_0$.

For the derivative $\dot{v}$, we obtain the expression

$$\dot{v} = -\frac{\mu}{c_0^2} g(t) \int_0^1 p(\xi, t) \int_0^\xi u^n(\eta, t) \, d\eta d\xi.$$

In this case the unperturbed system has the solution

$$\bar{u}(\xi, t) = \imath \sum_{k=1}^\infty \sin k\pi\xi [a_k e^{\imath k\pi t c_0} - a_{-k} e^{-\imath k\pi t c_0}],$$

$$\bar{p}(\xi, t) = -c_0 \sum_{k=1}^\infty \cos k\pi\xi [a_k e^{\imath k\pi t c_0} + a_{-k} e^{-\imath k\pi t} c_0 e^{-\imath k\pi t c_0}] - c_0 a_0.$$

Then the mean for the system (5.4) has the form

$$\psi = -\frac{1}{c_0^2} \lim_{T \to \infty} \frac{1}{T} \int_{t_0}^{t_0+T} g(t) \int_0^1 \bar{p}(\xi,t) \int_0^\xi \bar{u}^n(\eta,t) \, d\eta d\xi dt. \qquad (5.5)$$

It can be seen from the expression (5.5) that if $g(t) = g_0 e^{i\nu t}$, then resonances are possible with frequencies which are multiples of $c_0 \pi$. Indeed, if we write the expression for the mean in more details, then

$$\psi = \frac{-g}{c_0^2} \lim_{T \to \infty} \frac{1}{T} \int_{t_0}^{t_0+T} e^{-i\nu t} \times$$

$$\times \int_0^1 [-c_0 \sum_{k=1}^\infty \cos k\pi\xi(a_k e^{ik\pi t c_0} + a_{-k} e^{-ik\pi t} c_0 e^{-ik\pi t c_0}) - c_0 a_0] \times$$

$$\times \int_0^\xi [i \sum_{k=1}^\infty \sin k\pi\xi(a_k e^{ik\pi t c_0} - a_{-k} e^{-ik\pi t c_0})]^n \, d\eta d\xi dt.$$

In conclusion, we consider a more general hyperbolic system with small nonlinearities. We consider it from the point of view of applying Theorems 8.1 to 8.4 to investigate the stability of the trivial solution.

Let us write down the linear system

$$\frac{\partial u}{\partial t} = A(t)u = \sum_{i=1}^n A_i(\xi,t)\frac{\partial u}{\partial \xi_i} + B(\xi,t)u, \qquad (5.6)$$

where $\dim u = m$, and where the elements of the $m \times m$ matrices $A_i(\xi,t)$ and $B(\xi,t)$, are sufficiently smooth functions defined in the bounded region G with the smooth boundary Γ. The matrices $A_i(\xi,t)$ are assumed to be real and symmetric, continuous in the variable t and bounded with respect to the norm in $\mathcal{L}_2(G)$, $G \subset L^n$. Integrating by parts, we obtain

$$(Au, u) = \int_G \left(\sum_{i=1}^n A_i(\xi,t)\frac{\partial u}{\partial \xi_i} + B(\xi,t)u, u \right)_m d\xi =$$

$$= -\int_G \left(u, \sum_{i=1}^n A_i(\xi,t)\frac{\partial u}{\partial \xi_i} \right)_m d\xi + + \int_G \left(\left(B - \sum_{i=1}^n \frac{\partial A_i}{\partial \xi_i}\right) u, u \right)_m d\xi$$

$$+ \int_\Gamma \left(\sum_{i=1}^n A_i n_i u, u \right)_m ds,$$

where n_i are coordinates of the external normal to the boundary Γ. Hence,

$$\text{Re}(Au, u) = \int_G \left(\left(B + B^* - \sum_{i=1}^n \frac{\partial A_i}{\partial \xi_i} \right) u, u \right)_m d\xi +$$

$$+ \int_\Gamma \left(\sum_{i=1}^n A_i n_i u, u \right)_m ds.$$

We set the boundary conditions in the form

$$(u, \omega_j)|_\Gamma = 0, \quad j = \overline{1, r},$$

where $\omega_j(\xi)$ are continuous vector fields on the boundary Γ ($\partial G = \Gamma$). These conditions mean that the vector $u(\xi)$ for $\xi \in \Gamma$ must belong to some subspace $N(\xi)$ of the entire m-dimensional space, the subspace $N(\xi)$ being orthogonal to the vectors ω_j.

Let us assume that the rank of the matrix $\bar{A} = \sum_{i=1}^n A_i(\xi) u_i(\xi)$ does not change along the boundary Γ and that the subspace $N(\xi)$ is the maximum subspace on which $(Au, u)_m \leq 0$.

Suppose that the above assumption about the matrices $A_i(\xi, t)$, $B(\xi, t)$ are valid; let the differential operator $A(t)$ be defined in the space $\mathcal{L}^2(G)$ of functions with square integrable first derivatives and satisfying the boundary condition $u(\xi)|_\Gamma \in N(\xi)$, then, by virtue of the assumptions, the Cauchy problem $u(0, \xi) = u_0(\xi) \in D(A)$ is correctly posed in the space $\mathcal{L}^2(G)$.

In addition, under the conditions

$$\int_\Gamma \left(\sum_{i=1}^n A_i n_i u, u \right)_m ds \leq 0,$$

$$B + B^* - \sum_{i=1}^n \frac{\partial A_i}{\partial \xi_i} \leq 0$$

the operator $A(t)$ admits extension up to the maximal dissipative operator.

Let us consider the nonlinear system

$$\frac{\partial u}{\partial t} = \sum_{i=1}^{n} A_i(\xi,t)\frac{\partial u}{\partial \xi_i} + B(\xi,t)u + \mu F(t,u) \qquad (5.7)$$

satisfying the same conditions.

For the system (5.7) we choose the functional in the form

$$v = v_0(t,u) = \int (u,u)_m d\xi.$$

In this case we get the following expression for the derivative

$$\dot{v}_0 = \int_G \left(\sum_{i=1}^{n} A_i(\xi,t)\frac{\partial u}{\partial \xi_i} + B(\xi,t)u, u \right)_m d\xi + \mu \int_G (F(t,u),u)_m d\xi +$$

$$+ \int_G \left(u, \sum_{i=1}^{n} A_i(\xi,t)\frac{\partial u}{\partial \xi_i} + B(\xi,t)u \right)_m d\xi + \mu \int_G (u,F(t,u))_m d\xi.$$

The integrals which do not contain the small parameter μ are estimated by using the dissipativity of the operator $A(t)$. Given the solution $\bar{u}(\xi,t)$ of the unperturbed system (5.6), and using the theorems of Chapter 8, conditions of stability and instability of the trivial solution may be obtained for various forms of the function $F(t,u)$ (e.g. $F(t,u) = f(t)u^l$, $l > 0$ or $f(t)u^2$)

As we have noted already, known integrals may be used as the functional v_0 in various physical problems. The proposed technique makes it possible to develop a theory of parametric resonance for systems of equations of mathematical physics, a theory which will be valid not only when the system parameters are dependent on time periodically. Thus, the possibility arises of investigating the stability of solutions of equations with nonlinear differential operators with respect to small perturbations, when the integral of energy and some information on the solution of the nonperturbed equation are known.

In Ref. [77], we studied stabilization of the solution of the equation of vibration of a string under perturbation containing $\mathrm{sign}u_t \cdot u^{2l}$, i.e. accounting for the sign of velocity. The integral of energy was used.

8.6   Stability of Nonlinear Evolutionary Differential Equation with Perturbation

The above - described technique for analyzing equations with small non-linearity may be extended to nonlinear systems for which a functional with a constant - sign derivative is known. Fully integrable Hamiltonian systems possess this property. As an example, we will consider evolution and stability of solutions of nonlinear equations of the Boussinesque type under perturbations caused by interaction of the liquid with body surface having different elasticity and viscosity [51], [76], [77].

1. The basic facts of the Hamiltonian formalism described in Ref. [24] are the following:

a) The phase space, which is a $2m$-dimensional manifold $\mathcal{M}_{2n}$; a closed nondegenerate symplectic 2-form on the manifold $\mathcal{M}_{2n}$, which in local coordinates ξ has the form

$$\Omega = \sum_{\alpha,\nu=1}^{2n} \Omega_{\alpha,\nu} d\xi_\alpha \wedge d\xi_\nu,$$

where $\Omega_{\alpha,\nu}$ is a non – degenerate antisymmetric matrix satisfying the condition $\partial_\mu \Omega_{\alpha,\nu}$ + cyclic permutation = 0;

b) The Hamiltonian $\mathcal{H}$ on $\mathcal{M}_{2n}$. Here the equations of motion have the form $\dot{\xi}^\alpha = \sum_\nu \Omega^{\alpha,\nu} \dfrac{\partial \mathcal{H}}{\partial \xi^\nu}$, or, $\dot{\xi} = \{\xi, \mathcal{H}\}$; $\sum_\mu \Omega^{\alpha,\mu} \Omega_{\mu,\nu} = \delta^\alpha_\nu$, where

$$\{v(\xi), w(\xi)\} = \sum_{\alpha,\nu} \Omega^{\alpha,\nu} \frac{\partial v}{\partial \xi^\alpha} \frac{\partial w}{\partial \xi^\nu}$$

is the Poisson bracket. The evident generalization of finite – dimensional Hamiltonian systems is provided by the infinite – dimensional systems whose variables ξ are functions from some functional space B.

An example of the infinite-dimensional Hamiltonian system is the Sine-Gordon equation, whose phase space is the space of pairs of functions, $(u(x), s(x))$ decreasing at infinity. The Hamiltonian $\mathcal{H}$ and the form Ω in this case are ([102], pp . 269-382):

$$\mathcal{H} = \int_{-\infty}^{\infty} \left[\frac{1}{2}\gamma s^2 + \frac{1}{2}\gamma^{-1}(u_x^2 + 2m^2(1 - \cos u)) \right] dx,$$

$$\Omega = \int_{-\infty}^{\infty} \delta s(x) \wedge \delta u(x) dx$$

where m is the mass and γ is the interaction constant.

The equations of motion

$$\dot{u} = \gamma s, \quad \dot{s}\gamma^{-1} = (u_{xx} - m^2 \sin u)$$

are equivalent to the Sine-Gordon equation:

$$u_{tt} - u_{xx} + m^2 \sin u = 0.$$

The Hamiltonian system has the integral $\mathcal{H}$ whose derivative equals) by virtue of the system; other functionals (or functions) are integrals of the Hamiltonian system, if they commute with the Hamiltonian with respect to the Poisson bracket generated by Ω which introduces a structure of the Lie algebra. Existence of the integrals makes it possible to apply the theory of functionals of the Lyapunov-Chetaev type to analyze behaviour of the solutions of the Hamiltonian system: if there exists an integral of motion $\mathcal{H}$, it is a typical example of a 'neutral type' system [62].

2. We will formulate a theorem on stability of 'neutral type' system under nonlinear perturbation. The theorem can be effectively used to analyze the behaviour of solutions Hamiltonian systems under perturbation.

In the region $S_t\{x, t : \| x \|_B < N, t \geq 0\}$, we consider the equations

$$\dot{x} = f(x, t) + \mu F, \tag{6.1}$$

$$\dot{x} = f(x, t), \quad (\mu = 0). \tag{6.2}$$

Here, $f(x, t)$ is an operator – valued function with the domain $D(f)$, $\overline{D(f)} = B$. The perturbation μF: $\forall t \in [0, \infty)$, $\mu f : B \to B$, $x(t)$ is an abstract function in the Banach space B. It is assumed that for $0 < \mu < \mu_0$ the Cauchy problem $x(0) = x_0$ for Eqs. (6.1) and (6.2) is correctly posed for any segment $[0, T]$, $T > 0$; $x_0 \in D(A)$.

By $M_t(B)$ we denote the set of functional $v_0(x, t)$ on elements of B, $\forall t$, $v_0 : B \to \mathbf{R}^1$ is continuous in S_t; in addition, $v_0(x, t)$ is assumed to be Frechet – differentiable in S_t and differentiable with respect to t. The set of the functionals $v_0 \in M_t(B)$, positive – definite in x and satisfying the condition $\forall \delta > 0$, $\exists \gamma(\delta) > 0$, $\| x \| < \gamma \Rightarrow |v_0| < \delta$ will be denoted by $\bar{\Delta}_t(B) \in M_t(B)$.

By $S_R(z)$ we denote a sphere of radius R with the centre at z: $S_R(z) = \{x : \| x - z \| < R\}$, by $R(l_1, l_2)$ we shall denote the annulus $R(l_1, l_2) = \{x : l_1 < \| x \| < l_2\}$

We consider a perturbed functional $v = v_0 + u$

$$v_0 \in \bar{\Delta}_t(B), \quad u = u(t, x, \mu, \epsilon),$$

where the functional u is defined and differentiable in the annulus $R(\rho, \epsilon)$.

We consider the derivative of the function v by virtue of Eq. (6.1)

$$\frac{dv}{dt} = \frac{\partial v_0}{\partial t} + \frac{\partial v_0}{\partial x} f(x, t) + \mu \frac{\partial v_0}{\partial x} F +$$
$$+ \frac{\partial u}{\partial t} + \frac{\partial u}{\partial x} f(x, t) + \mu \frac{\partial u}{\partial x} F.$$

In this expression we examine terms containing F and u, single out the factor $\theta(\mu)$, depending on μ, and introduce the notation

$$\theta(\mu) k(x, t, \mu) = \mu \frac{\partial v_0}{\partial x} F + \frac{\partial u}{\partial t} + \frac{\partial u}{\partial x} f(x, t) + \mu \frac{\partial u}{\partial x} F$$

to ensure $k(x, t, 0) \not\equiv 0$. We denote by $\dot{v}_0$ the derivative of the functional v_0 computed with the help of the unperturbed equation (6.2).

Let us take $v_0 \in \bar{\Delta}_t(B)$. Since the functional $v_0(x, t)$ is positive - definite in x, there exists a positive - definite function w, $v_0 \geq w$, which is independent of t (according to definition). We denote the manifold

$w(x) = c_0$ by $\Xi_w\{c_0\}$ and the moving manifold $v_0(x,t) = c_0$ by $\Xi_w^t\{c_0\}$. Let us introduce the following sets $(c_0 > 0)$: $\text{Int}(\Xi_w\{c_0\}) = \{x : w(x) \leq c_0\}$, $\text{Int}(\Xi_w^t\{c_0\}) = \{x, t : v_0(x,t) \leq c_0\}$. Let us fix $\bar{\epsilon} > 0$ $(\bar{\epsilon} < N)$ and choose $w_0 > 0$ such that $\text{Int}(\Xi_w\{c_0\}) \subset S_{\bar{\epsilon}}(0)$; we choose $\sigma > 0$ as to satisfy the condition $2\sigma < w_0$. Since $v_0 \in \bar{\Delta}_t(B)$, there exists $\eta > 0$ such that $S_\eta(0) \in \text{Int}(\Xi_w^t\{w_0\})$.

Theorem 8.5 *Suppose that the following conditions are fulfilled:*

a) The Cauchy problem for Eqs. (6.1), (6.2) is correctly stated and its solution is defined on $[0, \infty)$; $x_0 \in D(f)$, with $\overline{D(f)} = B$.

b) There exists a functional $v_0 \in \bar{\Delta}_t(B)$ whose derivative $\dot{v}_0$ is non positive in $S_t \cap D(f)$ by virtue of Eq. (6.2).

Suppose also that for any $\epsilon > 0$ $(\epsilon < \epsilon_0 < N)$ there exist w_0, σ_0, γ $(2\sigma_0 < w_0, \gamma = \gamma(w_0, \sigma_0))$ related as above, such that $\forall \rho > \gamma$ $(\rho < \epsilon)$, in the annulus $R(\rho, \epsilon)$, $t > 0$, functionals $u(x, t, \mu, \epsilon)$, $k(x, t, \mu, \epsilon)$ are defined and exist:

a') $\delta > 0$ and $l > 0$ such that if for $t \in [t_0, t_0 + T]$, $\bar{x} \in r(\rho, \epsilon)$ (where $\bar{x}(t)$ is a solution of Eq. (6.2)), and $T > l$, then uniformly with respect to $t_0 > 0$

$$\int_{t_0}^{t_0+T} k(\bar{x}(t), t, \mu, \epsilon)\, dt < -\delta T;$$

b') a summable function $\bar{F}(t)$ and a constant F_0 as well as a non-decreasing function $\chi_1(\alpha)$, $\lim_{\alpha \to 0} \chi_1(\alpha) = 0$, such that for $t > 0$, $x \in R(\rho, \epsilon) \cap D(f)$

$$|k(x', t) - k(x'', t)| \leq \chi_1(\| x' - x'' \|)\bar{F}(t),$$

$$\int_{t_1}^{t_2} \bar{F}(t)\, dt \leq \bar{F}_0(t_2 - t_1)$$

on any finite segment $[t_1, t_2]$, $\bar{x}(t)$ being a solution of the unperturbed equation (6.2);

c') a non-decreasing function $\chi_2(\mu)$, $\lim_{\mu \to 0} \chi_2(\mu) = 0$, such that

$$|u(x, t, \mu, \epsilon)| < \chi_2(\mu), \quad x \in R(\rho, \epsilon).$$

Then Eq. (6.1) is (x, μ)-stable, i.e. $\exists \eta(\epsilon), \mu_0(\epsilon) > 0$, $\mu < \mu_0(\epsilon)$, $x_0 \in S_\eta(0) \cap D(A) \Rightarrow x(x_0, t) \in S_\epsilon(0)$.

The proof is similar to that of Theorem 8.1 of the present chapter.

3. In order to study the effect of perturbation in Hamiltonian systems $\dot{\xi} = \{\xi, \mathcal{H}\} + \mu F$, Theorem 8.5 may be used effectively with $v_0 = \mathcal{H}$, $v_0 \in \Delta_t(B)$ or $v_0 \in M_t(B)$. As is well – known from the study of stability of some particular solution, it is possible, using a change of variable, to move on to the investigation of the stability of the trivial solution. In this way it is possible to examine how perturbations affect a particular solution. In [62] the technique was used for finite-dimensional Hamiltonian systems. The Cauchy problem for the Sine-Gordon equation and for the nonlinear Schrödinger equation, where

$$\mathcal{H} = \int_{-\infty}^{\infty} ((2m)^{-1} \bar{\psi}_x \psi_x + \gamma |\psi_k|^4) dx,$$

$$\omega = \operatorname{Im} \int_{-\infty}^{\infty} \delta \bar{\psi} \wedge \delta \psi dx,$$

is Lyapunov - stable, since $\mathcal{H}$ is positive-definite. In the Korteweg – de Vries (KdV) equation

$$\mathcal{H} = \int_{-\infty}^{\infty} \left(\frac{1}{2} u_x^2 + u^3 \right) dx,$$

$$\omega = \int \int_{-\infty}^{\infty} \delta u(x) \wedge \delta u'(y) dx dy, \quad \mathcal{H} \in M_t,$$

so the above technique can be applied to study how perturbations affect the system. Here the phase space is the space of functions decreasing as $|x| \to \infty$.

We consider the perturbed KdV equation

$$u_t + u u_x + u_{xxx} = k F(t, x, u, u_x, \ldots). \tag{6.3}$$

The unperturbed equation $k = 0$ was studied in detail, and various classes of solutions were singled out, both periodic and decreasing at infinity. Using the perturbation theory, the KdV equation was studied in Refs. [84], [102], [92].

4. We consider the motion of a liquid layer of the unperturbed thickness h over the surface of a solid body undergoing deformation, $\tilde{z} = -h - f(t, x, y, \eta)$ where $\eta(t, x, y) = H - h$, with $H = \tilde{H} - f$; $\tilde{H}$ is the total thickness of the liquid layer, with the coordinate z directed normally to the unperturbed liquid surface. We also assume that $f \ll \eta$. The motion is described by the Euler equations

$$
\begin{aligned}
&v_t + (v\nabla)v = -\rho^{-1}\nabla p, \\
&\mathrm{div}\, v = 0, \\
&v = v(t, x, y, z).
\end{aligned}
\tag{6.4}
$$

We introduce the Bernoulli potential $v = \nabla\Phi$ and write Eq. (6.4) and the kinematic and dynamic boundary conditions on the surface $z = \eta(t, x, y)$. We also write the condition that there is no leakage on the surface $z = -h - f$ using dimensionless coordinates $x = x'/l$, $y = y'/l$, $z = z'/h$, $t = c_0 t'/l$, $\Phi = \Phi'/(v_0 l)$, $\eta = c_0\eta'/(v_0 h)$, $f = f'/(v_0 h)$. Here c_0 is the linearized sound velocity, l is a typical linear scale of perturbations, v_0 is a typical amplitude of the perturbation velocity. We also introduce small parameters $\epsilon = h/l$, $\mu = v_0/c_0$ and assume that ϵ^2 and μ have the same order of smallness. Then

$$
\frac{\partial\Phi}{\partial t} + \frac{\mu}{2}(\nabla\Phi)^2 + \frac{\mu}{2\epsilon^2}\left(\frac{\partial\Phi}{\partial z}\right)^2 + \eta\big|_{z=\mu\eta} = 0,
\tag{6.5}
$$

$$
\frac{\partial^2\Phi}{\partial z^2} + \epsilon^2\Delta\Phi = 0,
\tag{6.6}
$$

$$
\frac{\partial\eta}{\partial t} + \mu(\nabla\eta\cdot\nabla\Phi) - \frac{1}{\epsilon^2}\frac{\partial\Phi}{\partial z}\bigg|_{z=\mu\eta} = 0,
\tag{6.7}
$$

$$
\frac{\partial\Phi}{\partial z} + \mu\epsilon^2(\nabla f\cdot\nabla\Phi) + \epsilon^2\frac{\partial f}{\partial t}\bigg|_{z=-1-\mu\eta} = 0.
\tag{6.8}
$$

Here ∇ is the 'nabla' operator and Δ is the Laplacian in the variables x, y. The expansion in small parameters will be carried out with accuracy up to terms of the second order in ϵ^2 and μ. Let

$$\Phi = \phi(t, x, y) + \sum_{k \geq 1} \left(\frac{\partial^k \Phi}{\partial z^k} \right) z^k \frac{1}{k!} \tag{6.9}$$

be the expansion in power of z of the potential Φ. Then, substituting (6.9) into (6.6), we obtain

$$\left(\frac{\partial^2 \Phi}{\partial z^2} \right)_0 = -\epsilon^2 \Delta \phi,$$

$$\left(\frac{\partial^3 \Phi}{\partial z^3} \right)_0 = -\epsilon^2 \Delta \left(\frac{\partial \Phi}{\partial z} \right)_0,$$

$$\left(\frac{\partial^4 \Phi}{\partial z^4} \right)_0 = -\epsilon^4 \Delta^2 \phi.$$

Till now $\dfrac{\partial \Phi}{\partial z}$ remains unknown. Further, substituting the obtained expressions into the relation (6.9) for Φ and into the derivatives

$$\frac{\partial \Phi}{\partial x} = \frac{\partial \phi}{\partial x} + \frac{\partial}{\partial x} \sum_{k \geq 1} z^k \frac{1}{k!} \left(\frac{\partial^k \Phi}{\partial z^k} \right)_0,$$

$$\frac{\partial \Phi}{\partial z} = \sum_{k \geq 0} z^k \frac{1}{k!} \left(\frac{\partial^{k+1} \Phi}{\partial z^{k+1}} \right)_0,$$

we replace $\dfrac{\partial \Phi}{\partial z}, \dfrac{\partial \Phi}{\partial x}, \dfrac{\partial \Phi}{\partial y}$ in Eq. (6.8) with the obtained expressions. As a result, we arrive at the following relation for $\left(\dfrac{\partial \Phi}{\partial z} \right)_0$:

$$\left(\frac{\partial \Phi}{\partial z} \right)_0 = -\epsilon^2 \Delta \phi - \mu \epsilon^2 f \Delta \phi +$$

$$+ \frac{\epsilon^2}{2} \Delta \left(\frac{\partial \Phi}{\partial z} \right)_0 + \epsilon^2 \mu f \Delta \left(\frac{\partial \Phi}{\partial z} \right)_0 + \frac{\epsilon^4}{6} \Delta^2 \phi -$$

$$- \frac{\epsilon^4}{24} \Delta^2 \left(\frac{\partial \Phi}{\partial z} \right)_0 + \mu \epsilon^2 \frac{\partial f'}{\partial x} \frac{\partial}{\partial x} \left(\frac{\partial \Phi}{\partial z} \right)_0 - \epsilon^2 \frac{\partial f}{\partial t} +$$

$$+ \mu \epsilon^2 \frac{\partial f}{\partial y} \frac{\partial}{\partial y} \left(\frac{\partial \Phi}{\partial z} \right)_0 - \mu \epsilon^2 (\nabla f \nabla \phi),$$

and by applying the successive approximation technique, we obtain, with an accuracy up to terms of second order in ϵ^2 and μ the following equality

$$\left(\frac{\partial \Phi}{\partial z}\right)_0 = -\epsilon^2 \Delta \phi - \epsilon^2 \frac{\partial f}{\partial t} -$$
$$-\frac{\epsilon^4}{2}\Delta\frac{\partial \phi}{\partial t} - \frac{\epsilon^4}{3}\Delta^2\phi - \mu\epsilon^2(\nabla f \cdot \nabla \phi). \qquad (6.10)$$

Taking the relation for $\left(\dfrac{\partial^2 \Phi}{\partial z^2}\right)_0$, $\left(\dfrac{\partial^3 \Phi}{\partial z^3}\right)_0$, $\left(\dfrac{\partial^4 \Phi}{\partial z^4}\right)_0$ into account, we obtain the final expansion for Φ. Next, we substitute this expression into Eqs. (6.5), (6.7) and obtain (retaining the terms of first order in ϵ^2 and μ)

$$\frac{\partial \eta}{\partial t} + \mu(\nabla \eta \cdot \nabla \phi) + \Delta \phi + \frac{\epsilon^2}{3}\Delta^2\phi =$$
$$= -\frac{\partial f}{\partial t} - \frac{\epsilon^2}{2}\Delta\frac{\partial f}{\partial t} - \mu(\nabla f \cdot \nabla \phi), \qquad (6.11)$$
$$\frac{\partial \phi}{\partial t} + \frac{\mu}{2}(\nabla \phi)^2 + \eta = 0.$$

We note that the presence of the deformed layer is revealed mainly in the equation for the perturbation height, while in the equation for the velocity potential the corresponding terms enter with a higher order of smallness. Replacing the potential in Eq. (6.11) by using the formula $\psi = \phi + \dfrac{\epsilon^2}{3}\Delta\phi$, and taking into account first - order terms, we obtain the equations for η and ψ ($c_0^2 = gh$)

$$\frac{\partial \eta}{\partial t} + \mu(\nabla \eta \cdot \nabla \phi) + \Delta \psi = -\mu(\nabla f \cdot \nabla \psi) - \frac{\partial}{\partial t}\left(f + \frac{\epsilon^2}{2}\Delta f\right)$$
$$\frac{\partial \psi}{\partial t} + \frac{\mu}{2}(\nabla \psi)^2 + + \frac{\epsilon^2}{3}\Delta\eta + \eta = 0.$$

Returning to dimensionful variables and setting $\psi = v_0 l\psi - c_0^2 t$, we obtain the perturbed Boussinesque equations for the total height H and the plane velocity $v = \nabla\psi(t, x,, y) = (\psi_x, \psi_y)$:

$$H_t + (\nabla \cdot Hv) = -(\nabla \cdot f^{(H)}v) - (f^{(H)} + h^2 \Delta f^{(H)})_t,$$

$$v_t + \frac{1}{2}\nabla v^2 + g\nabla H + \frac{gh^2}{3}\nabla\Delta H = 0. \tag{6.12}$$

Here $f^{(H)}$ is the function f with $H - h$ in place of η.

An interesting case is the one in which the layer is visco – elastic with the function f of the form

$$f = G^{-1}(\eta - K \int_0^t \Gamma(t - \tau)\eta(\tau, x, y)\, d\tau), \tag{6.13}$$

where G is a dimensionless elasticity coefficient, K is a combined viscosity, and $\Gamma(t - \tau)$ is the relaxation kernel.

Passing in Eq. (6.12) to dimensionless variables with $v' = v/v_0$ and setting $f \to f'$, $\eta \to \eta'$ in Eq. (6-13), we get

$$\frac{\partial v'}{\partial t'} + \mu v'\frac{\partial v'}{\partial x'} + \frac{\epsilon^2}{3}\frac{\partial^3\eta}{\partial x'^3} + \frac{\partial\eta}{\partial x'} = 0,$$

$$\frac{\partial\eta'}{\partial t'} + \frac{\partial v'}{\partial x'} + \mu\frac{\partial}{\partial x'}(\eta', v') = -\mu(f', v')_{x'} - (f' + \epsilon^2 f'_{x',x'})_{t'}. \tag{6.14}$$

Further, setting in (6.13) $G^{-1} = \theta$, $\theta \sim \epsilon^2, \mu$, $f' = \theta s'$ we look for η' in (6.14) in the form

$$\eta' = v' + \frac{1}{4}\mu(v')^2 + \epsilon^2\alpha + \theta\beta + \mu\gamma$$

(with an accuracy up to term of the first order of smallness in ϵ^2, μ, and θ). Substituting the latter expression into (6.15), and grouping with required accuracy terms with corresponding power of the small parameters, we obtain $\alpha = -\frac{1}{6}\frac{\partial^2 v'}{\partial x'^2}$, $\gamma = 0$, $\beta = -\frac{1}{2}s'$. Substituting

$$\eta' = v' + \frac{1}{4}\mu(v')^2 - \frac{\epsilon^2}{6}(v')^2_{x'x'} - \frac{1}{2}\theta s'$$

into Eq. (6.14), and returning to dimensionful variables, we get with the same accuracy

$$v_t + \left(c_0 + \frac{3}{2}v\right)v_x + \frac{c_0 h^2}{6}v_{xxx} =$$

$$= \frac{c_0}{2G}\left(v_x - K\int_0^t \Gamma(t-\tau)v_x\, d\tau\right). \tag{6.15}$$

Having made the subtitution $c_0 + \frac{3}{2}v = u$ in Eq. (6.15), and assuming that the value of $\frac{c_0 h^2}{c}$ equals 1, we obtain the perturbed KdV equation (with the form of perturbation dictated by the actual model)

$$u_t + uu_x + u_{xxx} = \frac{c_0}{2G}\left(u_x - K\int_0^t \Gamma(t-\tau)u_x\, d\tau\right). \tag{6.16}$$

It is assumed here that the main direction in which perturbations propagate is the axis x, $u = u(t,x)$. If necessary, perturbations of higher orders may be constructed, and capillary effect may also be taken into account.

To study how the solution of the form

$$\bar{u} = 3A^2 \text{sech}^2\left[\frac{A}{2}(x - A^2 t)\right]$$

evolves under influence of perturbations of the indicated type, we diferentiate the integral

$$V = \frac{1}{2}\int_{-\infty}^{\infty} u^2(x)dx,$$

use the perturbed equation (6.16), and take this derivative for a solution with varied amplitude and phase. We obtain the equation (if $\Gamma = k_1 + k_2 \exp(-k_3(t-\tau))$

$$3\gamma_0 AA_t = \frac{c_0 K}{G}\int_{-\infty}^{\infty} \text{sech}^2\xi \int_0^t A^3(\tau)(k_1 + k_2 \exp(-k_3(t-\tau))) \times$$

$$\times \text{sech}^2 R(A,t,\tau,\xi)\tanh R(A,t,\tau,\xi)\, d\tau d\xi, \tag{6.17}$$

$$\gamma_0 = \int_{-\infty}^{\infty} \text{sech}^4\xi\, d\xi, \quad R = \frac{A(\tau)}{2}\left(\frac{2\xi}{A(\tau)} + A^2(t)t - A^2(\tau)\tau\right).$$

The solution of the indicated type, according to Eq. (6.17), will be damping, as in the case of the Boussinesque equations with perturbations, and the damping parameters are determined by the coefficients G, K and by the kernel $\Gamma(t - \tau)$.

5. Nonlinear completely integrable evolutionary equations with perturbation have been studied in a number of works. Various types of perturbation have been studied, e.g. θu and $\theta F(u)$, the KdV equation with variable coefficients, including the case of the small parameter at the highest derivative [84]. To study the evolution and stability of solutions, a number of efficient methods have been applied, such as the inverse scattering method, variation of a constant or direct expansion of the solution in a small parameter. Using the solutions described in this work, we can study evolution of solutions and draw conclusions on their stability or instability, provided that we know the integral of the nonperturbed equation or the Lyapunov functional, and also a certain class of solutions of the nonperturbed equation. It is concluded that the system is stable when the mean is negative, while instability occurs in the opposite case. The above – mentioned method is also applicable in the case of a fairly complex non – autonomous dependence of perturbations on t and x.

It should be noted that there is a certain link between soliton - like solutions and the Langmuir turbulence theory: this model reflects the fact that in the case of any current different from the laminary current, there exist actual non – homogeneities which propagate along the body surface along which the liquid flows. The existence of a viscoelastic layer on the interaction surface results in dissipation of perturbations and inhomogeneities; if the perturbed KdV equation is taken as a model, then the above-mentioned mean has the form

$$\psi(t_0) =$$
$$= -\frac{c_0 K}{2G} \lim_{T \to \infty} \frac{1}{T} \int_{t_0}^{t_0+T} \int_{-\infty}^{\infty} \bar{u}(t, x) \int_0^t \Gamma(t - \tau) \bar{u}_x(\tau, x) d\tau \, dx \, dt.$$

If the value and sign of the mean are controlled, it is possible to obtain the necessary values of the parameters K, G, and $\Gamma(t - \tau)$ which determine the dissipation.

A number of concepts is known which are intended to explain the nature of turbulent flows and the loss of laminarity (see, e.g. [103]). It is important that complex currents have different kind of inhomogeneities. If, on the other hand, the surface of interaction with the liquid flow is covered with a viscoelastic layer, then the latter will cause dissipation and excitation damping which often obstruct the laminar flow.

Reference [76] deals with the interaction of an incompressible liquid flow with body surfaces having viscoelastic properties, and with surfaces that react to the flow inhomogeneity by a specific change in the surface form. The optimization problem for determining the form of surface that realizes the fastest damping of isolated inhomogeneities of the liquid flow is solved. In this way it is established that an advance resistance of the body surface is brought about by a special deformation of isolated perturbations propagating along the surface. It has been established that the skin of dolphins and other cetaceans possesses similar properties due to the highly organized skin innervation, blood circulation and muscular systems [101]. In this case various inhomogeneities, including turbulent ones, are damped, the hydrodynamic resistance is diminished several times and, as a result, the flow becomes almost laminary even for high flow velocities.

The above – mentioned physical laws can be used as a basis for designing artificial surfaces and bodies possessing the ability to efficiently damp perturbations, and to diminish hydrodynamic resistance.

CHAPTER 9

Stability of Stable System Influenced by Small Random Perturbations

In Chapters 1 – 8, we considered deterministic systems containing small perturbations. However, there is another class of problems that offers considerable interest: namely problems with random features which are described by systems with random perturbations. The obvious way to construct a model for such a process is to regard the random factors as small additions to the deterministic RHS of the original system. The simplest situation to investigate stability is when the deterministic system has an asymptotically stable position of equilibrium. This question has been studied in detail in the literature [35]. In this approach, general conditions of smallness are imposed on perturbations in one way or another.

The present chapter is devoted to investigating stability in a neutral situation, when the unperturbed deterministic system has a stable equilibrium position, and specific conditions are imposed on random perturbations.

Suppose that the time evolution of some object under the action of random perturbations is described by a system of stochastic equations, having a small parameter. The unperturbed system will be assumed to consist of ordinary differential equations having stable stationary points in a part of the variables. In the described situation the solutions can, generally speaking, merge from any neighbourhood of the equilibrium position, even if it is stable with respect to relative perturbations of the initial data in the strictest sense. A functional of action was introduced in Ref. [105] for certain classes of random perturbations. Using this functional, it was shown that when the stationary point of an unperturbed system is uniformly asymptotically stable, then the mean time of exit increases exponentially as the small parameter decreases.

In the present chapter, we use the perturbed Lyapunov function technique to estimate the asymptotic time interval over which the solutions of the perturbed system, close to the stationary point of the non-perturbed system at the initial moment, will remain with a given probability in a fixed neighbourhood of the stationary point.

9.1 Construction of Perturbations of Lyapunov Functions under Small Random Perturbations

1. We consider the system of the Ito stochastic equations

$$dz = [f(t,z) + \mu F(t,z,\eta,\mu)]dt + \sqrt{\mu}\sigma(t,z,\eta,\mu)dw(t),$$

$$(1.1)$$

$$d\eta = b(t,\eta,\mu)dt + A(t,\eta,\mu)dw(t).$$

Here $\mu > 0$ is a small parameter; $z = (x,y)$, f and F are vectors with $n_1 + n_2$ components $(n_1 \geq 1)$; η and b are vectors with n_3 components; $w(t)$ is a standard Wiener process with values in $\mathbf{R}_{n_4}$ defined on a probabilistic space (Ω, S, P); σ and A are matrices mapping $\mathbf{R}_{n_4}$ into $\mathbf{R}_{n_1+n_2}$ and $\mathbf{R}_{n_3}$, respectively; $\sigma \cdot A^* = 0$.

We consider the region $U_H :\| x \|< H$, $y \in Q \cup \mathbf{R}_{n_2}$, $\eta \in \mathbf{R}_{n_3}$, $0 < \mu < 1$, $t > 0$, where Q is the variation range of the variables y.

Let us assume that in $\bar{U}_H$ the coefficients of the system (1.1) are continuous in all variables, and satisfy the Lipschitz condition in the variables z, η with the constant λ_1. By $N(t_0)$ we denote the set of random vectors (z_0, η_0), independent of the process $w_i(t) - w_i(t_0)$ $(1 \leq i \leq n_4, t \geq t_0 \geq 0)$. With regards to the unperturbed system

$$\frac{d\bar{z}}{dt} = f(t, \bar{z}) \tag{1.2}$$

we assume that it has a stationary point in the variables x, $x = 0$, i.e. $f_i(t, 0, y) \equiv 0$ for $1 \leq i \leq n_1$.

Let us introduce the matrices $C = (c_{ij}) = \sigma\sigma^*$ and $\mathcal{D} = (d_{ij}) = AA^*$. By L_μ we denote the generating differential operator of the process (t, z, η):

$$L_\mu = \frac{\partial}{\partial t} + (f_k + \mu F_k)\frac{\partial}{\partial z_k} + b_k\frac{\partial}{\partial \eta_k} + \frac{1}{2}\mu c_{kl}\frac{\partial^2}{\partial z_k \partial z_l} +$$
$$+ \frac{1}{2}\mu d_{kl}\frac{\partial^2}{\partial z_k \partial z_l}.$$

Here the summation is made over the repeating indices, $C(U_H)$ is a class of functions $\phi(t, z, \eta, \mu)$, continuous in $\bar{U}_H$ in all the variables; $C_2(U_H)$ is a class of functions $\phi(t, z, \eta, \mu) \in C(U_H)$, continuously differentiable with respect to t and twice continuously differentiable with respect to z, η in U_H; K is a class of continuous functions $a : \mathbf{R}_1 \to \mathbf{R}_1$, positive and increasing for positive values of the argument; $\hat{K}$ is a class of functions $a(\alpha) \in K$ such that $a(0) = 0$, $\tilde{K}$ is a class of functions $a(\alpha) \in K$ such that $\lim_{\alpha \to \infty} a(\alpha) = \infty$, and $a(\alpha)$ is twice differentiable with respect to α.

2.

Theorem 9.1 *Suppose that the following conditions are fulfilled:*

a) For the system (1.2) there exists a Lyapunov function $v_0(t,z) \in C_2(U_H)$, positive – definite in x, admitting an infinitely small upper bound in x, and satisfying in U_H the inequality

$$\frac{\partial v_0}{\partial t} + \frac{\partial v_0}{\partial z_l} f_l \leq 0.$$

The functions

$$u(t,z,\eta,\mu) = u_1(t,z,\eta,\mu) + u_2(t,z,\eta,\mu) \in C_2(U_H)$$

and

$$\psi(t,z,\eta,\mu) = \psi_1(t,z,\eta,\mu) + \psi_2(t,z,\eta,\mu) \in C(U_H)$$

are defined in U_H and satisfy:

b) For a certain function $\theta(\alpha) \in \hat{K}$, $L_\mu(v_0 + u) \leq \theta(\mu)\psi$ in U_H.

c) There exist functions $0 \leq \check{\Psi}(\eta) \in C_2(U_H)$, $a(\alpha) \in K$, $\Theta_2(\alpha)$ and $\chi(\alpha) \in \hat{K}$ and a contant $\phi_0 > 0$ such that $U_H|U_1| \leq \chi(\mu)$, $\psi_1 \leq \phi_0$, $\check{\Psi}(\eta) \leq a(\| \eta \|)$ and $|\psi_2| + |u_2| \leq \chi(\Theta_2(\mu)\check{\Psi}(\eta))$.

(d) The following conditions are satisfied:

d') there exists a function $\Theta_1(\alpha) \in K$ such that in U_H, $L_\mu\check{\Psi}(\eta) \leq \Theta_1(\mu)$ and $\Theta_2(\mu)\Theta_1(\mu)/\Theta(\mu) < \phi_0$; or

d") there exist functions $\Theta_3(\alpha) \in K$ and $\chi_1(\alpha) \in \hat{K}$ such that $L_\mu\chi_1(\check{\Psi}(\eta)) \leq \Theta_3(\mu)$ in U_H and for any $\lambda > 0$ the equation

$$\lim_{\mu \to 0} \frac{\Theta_3(\mu)}{\Theta(\mu)} \chi_1\left(\frac{\chi^{-1}(\lambda)}{\Theta_2(\mu)}\right) = 0$$

s valid.

Then, if a random value $(z_0, \eta_0) \in N(t_0)$ is such that for some onstant $E > 0$ we have $M\check{\Psi}(\eta_0) < E$ (provided that condition (d') is ulfilled) or $M\chi_1(\check{\Psi}(\eta_0)) < E$ (provided that condition $(d")$ is fulfilled), hen for arbitrarily small $\epsilon_1 > 0$, $\epsilon_2 > 0$, there exist $\kappa > 0$, $\delta > 0$, and $t_0 > 0$ such that for all μ, $0 < \mu < \mu_0$, $T = \kappa/\Theta(\mu)\phi_0$, the condition $Mv_0(t_0, z_0) < \delta$ gives

$$P\left(\sup_{t_0 \leq t \leq t_0 + T} \| x(t, t_0, z_0, \eta_0, \mu) \| \geq \epsilon_1\right) \leq \epsilon_2.$$

3. Proof. Without loss of generality we assume that $t_0 = 0$. We fix $\lambda > 0$. Let $\hat{Q}_1(\mu)$ be a maximal open connected subset of the set $\{\eta : \chi(\Theta_2 \check{\Psi}(\eta) < \lambda\}$ and containing the point $\| \eta \| = 0$. By virtue of the inequality $\check{\Psi}(\eta) \leq a(\| \eta \|)$, the set $\hat{Q}_1$ is non - empty for sufficiently small $\mu > 0$. We introduce the open set $\hat{Q}_2(\mu) = \{t, z, \eta : \eta \in \hat{Q}_1, y \in Q, t < T, \| x \| < \epsilon_1\}$

We consider the solution $z(t)$, $\eta(t)$ of the system (1.1) with the initial data $(z_0, \eta_0) \in N(0)$. By τ_μ we denote the moment when the process $(t, z(t), \eta(t))$ leaves $\hat{Q}_2(\mu)$ for the first time. Let $s_t = \min(t, \tau_\mu)$

$$P(\sup_{0 \leq t \leq T} \| x(t) \| \geq \epsilon_1) \leq P(x(s_T) \notin \hat{Q}_2(\mu)) + P(\eta(s_T) \notin \hat{Q}_2(\mu)). \quad (1.3)$$

We choose a number $v_0^{\epsilon_1} > 0$ such that the moving equipotential surfaces $v_0(t, z) = v_0^{\epsilon_1}$ in the variables x lie inside the ϵ-neighbourhood of the origin. By virtue of condition (c)

$$\psi \leq \phi_0 + \lambda \quad \text{and} \quad |u| \leq \chi(\mu) + \lambda \quad (1.4)$$

in $\hat{Q}_2$. Let us introduce the function $W = v_0 + \chi(\mu) + \lambda + u + \Theta(\mu)(\phi_0 + \lambda)(T - t)$. By virtue of conditions (a), (b), and the inequalities (1.4), we have $W \geq v_0 \geq 0$ and $L_\mu W \leq 0$ in $\hat{Q}_2(\mu)$. Hence the random process

$$W(s_t) = W(s_t, z(s_t), \eta(s_t), \mu)$$

is a non – negative supermartingale. Combining the properties of non – negative supermartingales and the Lyapunov function, we obtain

$$P(x(s_T) \notin \hat{Q}_2(\mu)) \leq P(\sup_{t \geq 0} v_0(s_t, z(s_t)) \geq v_0^{\epsilon_1}) \leq P(\sup_{t \geq 0} W(s_T) \geq v_0^{\epsilon_1}$$
$$\leq (\delta + 2(\chi(\mu) + \lambda) + \kappa(\phi_0 + \lambda)/\phi_0)/v_0^{\epsilon_1}. \quad (1.5$$

In order to estimate $P(\eta(s_T) \notin \hat{Q}_2(\mu))$ we consider two cases.

Case 1. Condition (d') is fulfilled, i.e.

$$P(\eta(s_T) \notin \hat{Q}_2(\mu)) = P(\check{\Psi}(\eta(s_T)) \geq \chi^{-1}(\lambda)/\Theta_2(\mu)).$$

We introduce the function $W_1 = \check{\Psi}(\eta) + \Theta_1(\mu)(T - t)$. By virtue of the inequality $\check{\Psi} \geq 0$ and condition (d'), we have $W_1 \geq 0$ and $L_\mu W_1 \leq 0$ in $\hat{Q}_2(\mu)$. From (1.5), we obtain

$$P(\eta(s_T) \notin \hat{Q}_2(\mu)) \leq (\Theta_2(\mu)E + \kappa)/\chi^{-1}(\lambda). \tag{1.6}$$

Here we have taken into account the fact that $M\check{\Psi}(\eta_0) \leq E$ and

$$\Theta_2(\mu)\Theta_1(\mu)/\phi_0\Theta(\mu) \leq 1.$$

Combining (1.3), (1.5), (1.6) and setting $\lambda = \chi(\sqrt{\kappa})$, we have

$$P(\sup_{0 \leq t \leq T} \| x(t) \| \geq \epsilon_1) \leq [\delta/v_0^{\epsilon_1} + 2\chi(\sqrt{\kappa})/v_0^{\epsilon_1} + \sqrt{\kappa} +$$

$$\tag{1.7}$$

$$+\kappa(\phi_0 + \chi(\sqrt{\kappa}))/v_0^{\epsilon_1}\phi_0] + [2\chi(\sqrt{\kappa})/v_0^{\epsilon_1} + \Theta_2(\mu)E/\sqrt{\kappa}].$$

The first bracket in (1.7) can be made smaller than $\frac{1}{2}\epsilon_2$ by choosing suficiently small $\delta > 0$ and $\kappa > 0$, the second one can be made smaller than $\frac{1}{2}\epsilon_2$ by choosing suficiently small $\mu_0 > 0$.

Case 2. Suppose that condition (d") is fulfilled, i.e.

$$P(\eta(s_T) \in Q_2) = P(\chi_1(\check{\Psi}(\eta(s_T)) \geq \chi_1(\chi^{-1}(\lambda)/\Theta_2(\mu))).$$

n this case, the function $W_2(s_t)$, where $W_2 = \chi_1(\check{\Psi}(\eta) + \Theta_3(\mu)(T - t)$ ossesses the properties of a non – negative supermartingale.

4. Suppose that elements of the matrices C, D and vectors B, E an be expanded in series in powers of μ or represented as sums of erms containing different powers of μ. Then the operator L_μ can be ormally written as

$$L_\mu = L_0 + \mu \sum_{l=1}^{\infty} \mu^{l-1}L_l.$$

A perturbation u of the Lyapunov function containing the infor- nation about the perturbing term of Eq. (1.1) may be represented s

$$u = \mu \sum_{l=1}^{k} \mu^{l-1}\hat{u}_l,$$

$\hat{u}_l$ being the solution of the recurrent system

$$L_0 \hat{u}_l = -\phi_{l-1}, \quad l = 1, 2, \ldots, k \tag{1.8}$$

where $\phi_{l-1} = \sum_{m=0}^{l-1} L_{l-m} \hat{u}_m - \hat{\phi}_{l-1}$, $\hat{\phi}_{l-1} \leq 0$, $\hat{u}_0 = v_0$.

5. Theorem 9.1 is interesting for the case in which $u_2 = \psi_2 \equiv 0$. In fact, this means that elements of the matrix C and the vector F are bounded functions of the variables z, η, t in U_H. If these are assumed to be unbounded, this means, on the other hand, that the range of problems considered may be extended, but, on the other hand, this will require certain conditions ensuring a sufficiently slow rate of evolution of the functions u and ψ along the trajectories of the system (1.1) (conditions (d') and (d")) and ensuring that some moments of the distribution of the random value η_0 are finite.

6.

Theorem 9.2 *Suppose that the following conditions are fulfilled:*

a) There exists a non-negative function $v_0(t, z) \in C_2(U_H)$ such that for a constant $E_1 > 0$, we have in U_H $v_0 \leq E_1$ and

$$\frac{\partial v_0}{\partial t} + \left(\frac{\partial v_0}{\partial z_l} \right) f_l \geq 0.$$

In U_H we define a function

$$u(t, z, \eta, \mu) = u_1(t, z, \eta, \mu) + u_2(t, z, \eta, \mu) \in C_2(U_H)$$

and a constant $\phi_0 > 0$ such that:

b) For a function $\Theta(\mu) \in \hat{K}$, we have $L_\mu(v_0 + u) \geq \theta(\mu)\phi_0$ in U_H.

c) For the function u condition (c) of Theorem 9.1 is fulfilled.

d) Condition (d") of Theorem 9.1 is fulfilled at $\lambda = 1$.

Then, if the random value $(z_0, \eta_0) \in N(t_0)$ is such that for a constant $E > 0$ we have $Mv_0(t_0, z_0) \leq E$ and $M\chi_1(\check{\Psi}(\eta_0)) \leq E$, it follows that

*for any $\epsilon_1 > 0$ ($\epsilon_1 < H$) and for any $\epsilon_2 > 0$ there exist $\kappa > 0$, $\mu_0 > 0$
such that for all μ, $0 < \mu < \mu_0$, $T = \kappa/\phi_0\Theta(\mu)$ we have*

$$P(\sup_{t_0 \leq t \leq t_0 + T} \| x(t, t_0, z_0, \eta_0, \mu) \| \geq \epsilon_1) \geq 1 - \epsilon_2.$$

7. Proof. Suppose that $t_0 = 0$. For $\lambda = 1$, we consider the set
$\hat{Q}_2(\mu)$ defined in the proof of Theorem 9.1:

$$P(\sup_{0 \leq t \leq T} \| x(t) \| < \epsilon_1) \leq P(x(s_T) \notin \hat{Q}_2(\mu)) + P(s_T \geq T). \quad (1.9)$$

The first term on the RHS of (1.9) is estimated as in the proof of
Theorem 9.1. To estimate the second term, we consider the function
$W_3 = v_0 + 1 + \chi(\mu) + u \geq 0$ in $\hat{Q}_2(\mu)$. From the Dynkin formula and
condition (b) we obtain $MW_3(s_T) \geq MW_3(0) + \theta(\mu)\phi_0 M s_t$. Hence,
$\theta(\mu)\phi_0 M s_T \leq E_3$, where $E_3 = E_1 + 2 + 2\chi(1)$. From the Chebyshev
inequality we have

$$P(s_T \geq T) \leq \frac{E_3}{\kappa}. \quad (1.10)$$

Finally,

$$P(\sup_{0 \leq t \leq T} \| x(t) \| < \epsilon_1) \leq$$
$$\leq [E|\kappa] + [(E + \kappa\theta_3(\mu)|\phi_0\theta(\mu)) | \chi_1(\chi^{-1}(1)|\theta_2(\mu))] \quad (1.11)$$

The RHS of (1.11) can be made smaller than ϵ_2 by choosing a sufficiently large $\kappa > 0$ and a sufficiently small $\mu_0 > 0$.

8. The theorems formulated above might have been formulated
without refering to the specific form of the system (1.1), but by using
the concept of a weak infinitesimal operator of a strictly Markov process. Using the latter approach, Theorems 9.1 and 9.2 can be extended
to the case in which the perturbing process is a Markov chain with a
finite (or countable) number of states, the processes with discrete time,
and so on.

9. For $0 < \epsilon_1 < H$, we introduce the random value

$$\check{\tau}_\mu = \inf(t \geq 0 :\| x(t_0 + t, t_0, z_0, \eta_0, \mu) \| \geq \epsilon_1).$$

Now we formulate, in terms of the probabilistic characteristics of $\check{r}_\mu$, a corollary from Theorem 9.1 and 9.2.

Theorem 9.3 *Suppose that the following conditions are fulfilled:*

a) There exists a Lyapunov function $v_0(t,z) \in C_2(U_H)$ of the system (1.2) which is positive - definite in x and which possesses an infinitely small upper bound in x , such that

$$\frac{\partial v_0}{\partial t} + \left(\frac{\partial v_0}{\partial z_l}\right) f_l \equiv 0 \quad in \ U_H.$$

In U_H a function $u = u_1 + u_2 \in C_2(U_H)$ and a constant $\phi_0 > 0$ are defined such that:

b) $L_\mu(v_0 + u) = \theta(\mu)\phi_0$ in U_H for a function $\theta \in \hat{K}$.

c) Condition (c) of Theorem 9.2 and condition (d") of Theorem 9.1 are fulfilled.

Then, if the random value $(z_0, \eta_0) \in N(t_0)$ is such that for a contant $E > 0$ we have $M\chi_1(\check{\Psi}(\eta_0)) < E$, it follows that for any $\epsilon_1 > 0$ ($\epsilon_1 < H$) and for any $\epsilon_2 > 0$ there exist $\delta > 0$, $\kappa_2 > \kappa_1 > 0$, $\mu_0 > 0$ such that for all μ, $0 < \mu < \mu_0$ $Mv_0(t_0, z_0) < \delta$ we have

$$P(k_1 < \Theta(\mu)\phi_0\check{r}_\mu \le k_2) \ge 1 - \epsilon_2.$$

10. Under conditions of Theorem 9.3 we write $\check{r}_\mu = O\left(\dfrac{1}{\Theta(\mu)\phi_0}\right)$ (P – almost everywhere). Let us consider stationary random processes $\eta_1(t)$, $\eta_2(t)$ with spectral densities $\Phi_1(\omega) = \dfrac{2\alpha\Omega^2}{\pi(4\alpha^2\omega^2 + \omega^2 - \Omega^2)^2)}$ and $\Phi_2(\omega) = \dfrac{\alpha^2}{\pi(\alpha^2 + \omega^2)}$. For $\alpha \ll \Omega$ almost all the energy of the process is concentrated in a narrow frequency range (of the order of α) near the point $\omega = \sqrt{\Omega^2 - 2\alpha^2} \approx \Omega$. Such a random process is called narrow - banded. The spectral density $\Phi_2(\omega)$ is monotonically decreasing for all $\omega > 0$. The more α increases, the more slowly $\Phi_2(\omega)$ decreases. Such

a random process is called wide - banded. We model the processes η_1 and η_2 as solution of the Ito equations

$$dn_1 = \Omega \eta_3 dt,$$
$$d\eta_3 = -(\Omega \eta_1 + 2\alpha \eta_3)dt + 2\sqrt{\alpha}dw(t), \tag{1.12}$$
$$\Omega > \alpha > 0, \quad M\eta(0)\eta_3(0) = 0, \quad M\eta_1^2(0) = M\eta_3^2(0) = 1,$$

$$d\eta_2 = -\alpha \eta_2 dt + \sqrt{2}\alpha dw(t),$$
$$M\eta_2(0) = 0, \quad M\eta_2^2(0) = \alpha. \tag{1.13}$$

Example 1. *Effect of a narrow – banded random process on a linear oscillator without damping.* Let us consider the equation

$$\ddot{x} + \omega^2 x = \mu \eta_1(t), \tag{1.14}$$

where η_1 is defined by the system (1.12), and the perturbed Lyapunov function

$$v = \frac{1}{2}(\omega^2 x^2 + \dot{x}^2) + \mu[(A_1 x + A_2 \dot{x})\eta_1 + (A_3 x + A_4 \dot{x})\eta_3] + $$
$$+\mu^2[C_1\eta_1^2 + C_2\eta_1\eta_3 + C_3\eta_3^2],$$

where the coefficients A_i, C_i are uniquely determined by (1.8), and

$$L_\mu v = \mu^2 \pi \Phi_1(\omega) > 0.$$

It is easy to show that for the functions $v = v_0 + u$, $a(\beta) = 2\beta^2$, $\Psi = \eta_1^2 + (\alpha/\Omega)\eta_1 + \eta_3^2$, $\chi(\beta) = C_8(\beta + \sqrt{\beta})$ (here $C_8 > 0$ is a constant), $\chi_1(\beta) = \beta^2$, all the conditions of Theorem 9.3 are fulfilled. We have obtained $\check{\tau}_\mu = O(1/\mu^2 \pi \Phi_1(\omega))$ under the condition that $M\eta_1^4(0) < \infty$, $M\eta_3^4(0) < \infty$. The function $\dfrac{1}{\mu^2 \pi \Phi_1(\omega)}$ has a local minimum at $\omega \approx \Omega$ (resonance). When $\omega \to \infty$, it increases rapidly (as ω^4).

Example 2. *Effect of a wide - banded random process on a linear oscillator without damping.* Consider the equation

$$\ddot{x} + \omega^2 x = \mu \eta_2(t), \tag{1.15}$$

where η_2 is defined by the system (1.13). In this example, we use the perturbed Lyapunov function

$$v = \frac{1}{2}(\omega^2 x^2 + \dot{x}^2) + \mu(A_1 x + A_2 \dot{x})\eta_2 + \mu^2 C \eta_2^2,$$

where $(\omega^2 + \alpha^2)A_1 = -\omega^2(\omega^2 + \alpha^2)A_2 = \alpha$, $2(\omega^2 + \alpha^2)C = 1$, $L_\mu v = \mu^2 \pi \Phi_2(\omega)$.

Using Theorem 9.3, we obtain $\check{\tau}_\mu = O(1/\mu^2 \pi \Phi_2(\omega))$. The function $1/\mu^2 \pi \Phi_2(\omega)$ increases monotonically for $\omega > 0$. When $\alpha \to \infty$, the spectral density $\Phi_2(\omega)$ tends to a constant in all frequencies. Let us replace the wide - banded process in Eq. (1.5) with the white noise of the corresponding intensity

$$\ddot{x} + \omega^2 x = \mu\sqrt{2}\dot{w}(t).$$

We use the function $v = \frac{1}{2}(\omega^2 x^2 + \dot{x}^2)$, $L_\mu v = \mu^2 > 0$. We obtain $\check{\tau}_\mu = O(1/\mu^2)$ Note that $\lim_{\alpha\to\infty} \mu^2 \pi \Phi_2(\omega) = \mu^2$.

9.2   Averaging in Some Stochastic Systems

In this section we examine the possibility of stabilizing solutions of a multi – frequency system on the integral curve of an averaged system by jamming the frequencies with processes of the white-noise type.

We will consider a multi – frequency system of ordinary differential equations which contains a controlling effect of the white-noise type:

$$\begin{cases} \dot{x} = \mu X(x,\phi), \\ \dot{\phi} = \omega(x) + \mu Y(x,\phi) + \sqrt{2}\sigma(\mu)\dot{w}(t). \end{cases} \tag{2.1}$$

Here $\dim x = n \geq 1$, $\dim \phi = m$, $X(x,\phi)$ and $Y(x,\phi)$ are periodic in ϕ with the period 2π, $\sigma(\mu)$ is a $(m \times l)$ - matrix, $w(t)$ is a standard Wiener process with values in $\mathbf{R}^l$ defined on a probabilistic space (Ω, S, P).

Let us introduce the system (2.2) obtained by the formal averaging of the equation for slow variables in (2.1) over the angles:

$$\dot{\xi} = \mu X_0(\xi), \quad X_0(x) = (2\pi)^{-m}\int_0^{2\pi} X(x,\phi)d\phi. \tag{2.2}$$

We denote the minimum eigenvalue of the matrix $\sigma\sigma^\perp$ by $\sigma_0(\mu)$. Let us expand the function $X(x,\phi)$ into the Fourier series

$$X(x,\phi) = X_0(x) + \sum_{\|k\|>0} \left(A_k(x)\cos(k,\phi) + B_k(x)\sin(k,\phi) \right).$$

We will formulate and prove a theorem on the proximity of solutions of the systems (2.1) and (2.2) emerging at the moment $t_0 \geq 0$ from the same point.

Theorem 9.4 *Suppose that the following conditions are fulfilled:*

a) The coefficients of (2.1) are continuous in all the variables, the vector functions $\omega(x)$, $A_k(x)$, $B_k(x)$ are continuously differentiable in the variables x in the region $x \in Q$, $\phi \in [0,2\pi)^m$ and for some constants C_1, $C_2 \geq 0$ in this region, we have

$$\| X_0(x_1) - X_0(x_2) \| \leq C_1 \| x_1 - x_2 \|,$$
$$\left\| \frac{\partial \omega}{\partial x} \right\| \leq C_2$$
$$\max(\| X_0 \|, \| X \|, \| Y \|, \| \omega \|) \leq C_1,$$
$$\sum_{\|k\|>0} \frac{1}{\| k \|} \frac{\| A_k \| + \| B_k \| + \| k \|}{\left\| \dfrac{\partial A_k}{\partial x} \right\| + \left\| \dfrac{\partial B_k}{\partial x} \right\|} \leq C_1.$$

b) $\sigma_0(\mu) > 0$ for $\mu > 0$ and $\lim_{\mu\to 0} \mu\sigma^{-2}(\mu) = 0$.

c) For a certain point $x_0 \in Q$ the solution $\xi(t)$ $(\xi(t_0) = x_0)$ of the system (2.2) is defined for all $t \geq t_0 \geq 0$ in region Q together with a ρ-neighbourhood $(\rho > 0)$.

Then, for any arbitrarily small $\epsilon > 0$ $(\epsilon < \rho)$ and any arbitrarily large > 0 we have

$$\lim_{\mu\to 0} P(\sup_{t_0 \leq t \leq t_0 + L/\mu} \| x(t) - \xi(t) \| \geq \epsilon) = 0,$$

where $x(t)$ is a component of the solution $(x(t),\phi(t))$ of the system (2.1) with the initial data $(x_0,\phi_0) \in N(t_0)$.

Proof. Let us introduce the variable $z = x - \xi$. We get the system

$$\begin{cases} \dot{z} = \mu(X(z + \xi, \phi) - X_0(\xi)), \\ \dot{\phi} = \omega(z + \xi) + \mu Y(z + \xi, \phi) + \sqrt{2}\sigma(\mu)\dot{w}(t), \\ \dot{\xi} = \mu X_0(\xi). \end{cases} \quad (2.3)$$

Using Theorem 9.1, we investigate the stability of the equilibrium position $z = 0$ of the system (2.3) on a time interval of the length L/μ in the region $\| z \| < \epsilon$, $\phi \in [0, 2\pi)^m$, $\xi \in Q$.

Let v_0 be a function positive – definite in z and twice continuously differentiable in the region $\| z \| < \epsilon$. We consider the perturbed Lyapunov function in the form

$$v = v_0 + \mu(\nabla v_0, u(z, \xi, \phi, \mu)),$$

where

$$u = \sum_{\|k\|>0} (C_k(z + \xi) \cos(k, \phi) + D_k(z + \xi) \sin(k, \phi)),$$

$$\Delta \cdot C_k(x) = (\sigma^k \cdot A_k(x) + (k, \omega(x))B_k(x)),$$

$$\Delta \cdot D_k(x) = ((k, \omega(x))A_k(x) + \sigma^k B_k(x)),$$

$$\Delta = (\sigma^k)^2 + (k, \omega(x))^2, \quad \sigma^k = (\sigma, \sigma^\perp)_{ij}k_i.$$

After some simple calculation, using condition (a) we find that in the considered region

$$L_\mu v \le \mu C_1 \| z \| \cdot \| \nabla v_0 \| + (\mu \sigma_0^{-1})^2 C_3 \left(\sigma_0 \| \frac{\partial^2 v_0}{\partial z^2} + \right.$$
$$\left. + (\sigma_0 + C_2) \| \nabla v_0 \| \right),$$

with a constant $C_3 \ge 0$.

Since $|(\nabla v_0, u)| \le C_1 \| \nabla v_0 \| \sigma_0^{-1}(\mu)$, for any function $v_0(z)$ positive – definite and twice continuously differentiable with respect to z, with $\| z \| \| \nabla v_0 \| \le C_5 v_0(z)$, all the conditions of Theorem 9.1. (here z plays a role of x, (ϕ, ξ) plays a role of y, the variables η are absent).

Let us take $v_0 = \| z \|^2$. Since the variables η are absent, we may take $W(\eta, \mu) = 0$, and since we are interested in the behaviour of the solution with the initial data $z_0 = 0$, δ can also be set equal to zero. Taking the remark after proof of Theorem 9.1 into account, we obtain the estimate

$$P(\sup_{t_0 \leq t \leq t_0 + L/\mu} \| z(t) \| \geq \epsilon) \leq \mu C_6 (\sigma_0(\mu))^{-2} (\sigma_0 + C_2) \epsilon^{-2}, \qquad (2.4)$$

and, since $z = x - \xi$, it follows from condition (b) and (1.11) that the theorem is valid.

Remark. For systems with constant frequencies, in condition (a) of Theorem 9.4, C_2 can be made equal to zero and, therefore, it is possible to replace condition (b) with a less retsrictive condition

$$\lim_{\mu \to 0} \mu \sigma_0^{-1}(\mu) = 0.$$

It is possible to present examples which show that the condition (2.5) is essential and may not be relaxed without additional assumptions on properties of combinative frequencies.

Let us assume that the averaged system (2.2) has an asymptotically stable equilibrium position $\xi = 0$; here the asymptotic stability is ensured by the negativeness of the real part of the eigenvalues of the matrix of the system of equations in variations of the averaged system. Then there exists a positive - definite quadratic form $v_0(\xi)$ whose derivative, by virtue of the system (2.2), is a function which is negative definite (at least, in a sufficiently small ϵ_1-neighbourhood of the point $\xi = 0$).

Let us introduce the random variable

$$\tau_\mu = \inf(t \geq t_0 : \| x(t, t_0, x_0, \phi_0, \mu, \omega) \| \geq \epsilon) - t_0$$

describing the time during which the solutions of the system (2.1) remain in the ϵ-neighbourhood of the equilibrium position $x = 0$ of the averaged system.

For some $T = T(\mu)$, we consider the set

$$Q_1 = \{t, x, \phi : t < t_0 + T, \parallel x \parallel < \epsilon, \ \phi \in [0, 2\pi)^m\}.$$

We choose a constant $c_1 > 0$ from the condition $|(\nabla v_0, u)| \leq C_1 \sigma_0^{-1}$ in Q_1. Then, in Q_1, we have

$$v = v_0(x) + \mu C_1 \sigma_0^{-1} + \mu(\nabla v_0, u(x, 0, \phi, \mu) \geq v_0(x).$$

We consider the function $W_2 = \exp(\mu^{-\alpha} v)$ for some α, $0 \leq \alpha < 1$ and estimate $L_\mu W_2$ in Q_1. We find that for a constant C_2 in Q_1

$$L_\mu W_2 \leq C_2(\mu^{1-\alpha}\sigma_0^{-1})^2 \exp(\mu^{-\alpha}(C_2\mu^{1-\alpha}\sigma_0^{-2})) = \chi_1(\mu).$$

Let us introduce the function

$$W_3 = W_2 + \chi_1(\mu)(T + t_0 - t).$$

It is obvious that $W_3 \geq W_2$ and $L_\nu W_3 \leq 0$ in Q_1. By τ_1 we denote the moment when the process $(t, x(t), \phi(t))$ leaves the region Q_1 for the first time. As in the proof of Theorem 9.4, we obtain the inequalities

$$P(\tau_\mu^\epsilon \leq T) = P(\sup_{t_0 \leq t \leq t_0 + T} \parallel x(t) \parallel \geq \epsilon) = P(\parallel x(\tau_1 \parallel \geq \epsilon) \leq$$
$$\leq P(v_0(x(\tau_1)) \geq w(\epsilon)) \leq P(W_3(\tau_1) \geq \exp(\mu^{-\alpha} w(\epsilon))) \leq$$
$$\leq M W_3(t_0) \exp(-\mu^{-\alpha} w(\epsilon)) \leq \tag{2.5}$$
$$M \exp(-\mu^{-\alpha}(w(\epsilon) - v_0(x_0 - \mu 2 C_1 \sigma_0^{-1})) + T\chi_1(\mu) \exp(-\mu^{-\alpha} w(\epsilon)))$$

We require that the inequality $v_0(x_0) < \frac{1}{2} w(\epsilon)$ holds with probability 1 and, having chosen $T = (\mu^{1-\alpha}\sigma_0^{-1})^{-2} \exp(C\mu^{-\alpha})$, with $C < w(\epsilon)$, we find that the RHS of (2.5) tends to zero as $\mu \to 0$, if the condition $\lim_{\mu \to 0} \mu^{1-\alpha}\sigma_0^{-2} = 0$ is fulfilled. We formulate this result as

Theorem 9.5 *Suppose that the following conditions are fulfilled:*

a) Condition (a) of Theorem 9.4 holds.

b) $\sigma_0(\mu) > 0$ *for* $\mu > 0$, *and for some* α, $0 \le \alpha < 1$

$$\lim_{\mu \to 0} \mu^{1-\alpha} \sigma_0^{-2}(\mu) = 0.$$

c) *The averaged system has an asymptotically stable equilibrium position* $x = 0 \in Q$, *where the asymptotic stability is ensured by negativeness of the real part of the eigenvalue of the matrix of the system of equations in variations of the averaged system.*

Then for any $\epsilon > 0$ *there exists a constant* $C > 0$ *such that*

$$\lim_{\mu \to 0} P(\tau_\mu^\epsilon > \mu^{-2(1-\alpha)} \sigma_0^2(\mu) \exp(C\mu^{-\alpha})) = 1.$$

provided that $P(v_0(x) \le \frac{1}{2} w(\epsilon)) = 1.$

This theorem shows that if the averaged system has an asymptotically stable position of equilibrium, then the solution of the initial system, which is close to the stationary point at the initial moment of time can be stabilized by jamming the frequencies in its neighbourhood during an exponentially large time interval for $\mu \to 0$.

Bibliography

[1] ABALKIN, V.K., AKSENOV, V.P., GREBENNIKOV, E.A., DEMIN, V. C., AND RYABOV, YU. A., A Reference Manual on Celestial Mechanics and Astrophysics, Moscow, Nauka, 1976. (in Russian)

[2] ANASHKIN, O.V., Asymptotic Stability in Non – Linear Systems and a Critical Case of Imaginary Roots in the Theory of Stability of Motion Diff. Eq., 14, No. 9, 1689 - 1691, (1978) (in Russian)

[3] ANASHKIN, O.V. AND HAPAEV, M.M., A Comparison Method and Investigation of Stability of Ordinary Differential Systems Containing Perturbations, ibid, 22, No. 9 1604 – 1606, (1986) (in Russian)

[4] ANASHKIN, O.V. AND HAPAEV, M.M., A Comparison Method and Investigation of Stability of Ordinary Differential Systems Containing Perturbations II, ibid, 25, No. 2 187 – 192, (1989) (in Russian)

[5] ANASHKIN, O.V., On Asymptotic Stability in Non – Linear Systems, ibid, 14, No. 8 1490 – 1493, (1978) (in Russian)

[6] ANASHKIN, O.V., On Investigation of Stability under Permanently Acting Perturbations in 'Neutral Case', ibid, 14, No. 6 1124 – 1127, (1978) (in Russian)

[7] ARNOLD, V.I., Applicability Conditions and Error Estimates of the Averaging for System which Pass through Resonances in their Evolution, Dokl. Akad. Nauk. SSSR, 161, No. 1, 9 –12 (1965) (in Russian)

[8] ARNOLD, V.I., Small Denominators and Problem of Stability of Motion in Classical and Celestial Mechanics, Usp. Math. Nauk, 18, 32 – 192 (1963) (in Russian)

[9] AZAROVA, O.A., KUZNETSOVA, I.V., AND HAPAYEV, M.M., Stability Investigations in Some Problems of Non – Linear Mechanics, Prikl. Math. Mech. 48, No. 2, 221 – 224 (1984) (in Russian)

[10] BELETSKII, V.V., Satellite Motion with Respect to the Mass Centre. Moscow, Nauka, 1965 (in Russian)

[11] BOGOLYUBOV, N.N. AND MITROPOLSKII, YU.A., Asymptotic Methods in the Theory of Non – Linear Oscillations. Moscow, Nauka, 1963 (in Russian)

[12] BOGOLYUBOV, N.N., MITROPOLSKII, YU.A. AND SAMOILEN-KO, A.M., Accelerated Convergence Method in Non – Linear Mechanics. Kiev, Naukova Dumka, 1969 (in Russian)

[13] BRUNO, A.D., The Analytic Form of Differential Equations, Trudy Moscow Math. Obshch., 27, 119 – 262 (in Russian)

[14] BRUNO, A.D., Normal Form and Averaging Method, Dokl. Akad. Nauk SSSR, 230, No. 2, 257 – 260 (1976)(in Russian)

[15] BRUNO, A.D., Normal Form of real Differential Equations, Math. Zametki, 18, No. 2, 227 – 241 (in Russian)

[16] BROWER D. AND CLEMENCE, J., Methods of Celestial Mechanics, Academic Press, 1961

[17] CHERNOUS'KO, F.L., Resonance Phenomena in Satellite Motion with Respect to the Mass Centre, Vich. Math. i Math. Phys. 3, No. 3, 528 – 538 (1963) (in Russian)

[18] CHETAYEV, N.G., Stability of Motion, Moscow – Leningrad, Gostekhiszdat, 1946 (in Russian)

[19] CORDUNIANU, C., Applications of Differential Inequalities in Stability Theory, An. Sci. Univ. 'Al I Cuza'; Iasi Sect. Mat. 6, 47 – 58 (1960)

[20] DARVIN, J.H., The Tides and Kindered Phenomena in Solar System, London, Murray, 1898

[21] DRIVER, R.D., Existence Theory for Delay – Differential Equations, Contr. to Diff. Eq. I, No. 3, 317 – 336 (1963)

[22] DUBOSHIN, G.N., Celestial Mechanics: Main Problems and Methods, Moscow, Nauka (1975) (in Russian)

[23] DUBOSHIN, G.N., Stability of Motion with Respect to Permanently Acting Perturbations. Trudy GAISh, 14, I, 153 – 164 (1940) (in Russian)

[24] DUBROVIN, B.A., NOVIKOV, S.P., FOMENKO, A.T., Modern Geometry, vols. I – III Springer Verlag, (Translated from Russian)

[25] FALIN, A.I., Investigation of Stability of Weakly Autonomous Systems by Averaging Methods, Diff. Urav., 16, No. 2, 252 – 257 (1980) (in Russian)

[26] FALIN, A.I., Investigation of Stability under Permanently Acting Perturbations in Special 'Neutral Case', ibid, 15, No. 12, 2278 – 2281 (1979) (in Russian)

[27] FILATOV, A.N., Asymptotic Methods in Theory of Differential and Integral Equations, Tashkent, Fan, 1974 (in Russian)

[28] FILATOV, A.N., Asymptotic Methods in Theory of Differential and Integro – Differential Equations, Tashkent, Fan, 1971 (in Russian)

[29] FILATOV, A.N., On Partial Averaging in Systems of Ordinary Differential Equations, Diff. Urav. 6, No. 6, 1118 – 1120 (1970) (in Russian)

[30] FILATOV, O.P., Stability of Multi – Frequency Systems under Resonance without Synchronization of Phase Variables, ibid, 19, No. 3, 543 – 546 (1983) (in Russian)

[31] FILIPPOV, A.F., Differential Equation with Discontinuous Right Side, Moscow, Nauka 1985 (in Russian)

[32] FOMIN, V.N., Mathematical Theory of Parametric Resonance in Linear Systems, Leningrad University Press, 1972 (in Russian)

[33] GODDUM, T.W., Inequalities and Quadratic Forms, Pacific T. of Math., 8, No. 3, 411 – 414 (1958)

[34] GERMANIDZE, V.E. AND KRASOVSKII, N.N., On Stability under Permanently Acting Perturbations, Prikl. Math. Mech., 21, No. 6, 769 – 774 (1957) (in Russian)

[35] GIKHMAN, I.I. AND SKOROKHOD, A.V., Introduction to the Theory of random Processes, Moscow, Nauka 1977 (in Russian)

[36] GOLDREICH, D., History of the Lunar Orbit, Rev. Geophys., 4, No. 4, 411 (1966)

[37] GOLDREICH, D., On the Eccentricity of Satellite Orbits in the Solar System, Mon. Not. Roy. Astron. Soc., 126, 257 – 268 (1963)

[38] GOLUBEV, V.G., On regions of Impossible Motion in the Three - Body Problem, Sov. Dokl., 174, No. 4, 761 - 770, (1967) (in Russian)

[39] GOLUBEV, V.G., On Upper estimates of Distances Between Bodies in the Unbounded Three - Body Problem, Lett. Astron. Mag., 3, No. 2, 82 - 85 (1977) (in Russian)

[40] HALANAY, A., An Averaging Method for Systems of Differential Equations with Delayed Argument, Rev. math. Pures Appl., 4, No 3, 467 - 483 (1959)

[41] HALE, J.K., Functional Differential Equations, Academic Press 1971

[42] HAPAYEV, M.M. AND ANASHKIN, O.V., On Stability Investigation in Systems of Ordinary Differential Equations with Almost - Periodic Coefficients, Dokl. Akad. Nauk SSSR, 240, No. 5, 1028 - 1031 (1978) (in Russian)

[43] HAPAYEV, M.M., Generalization of the Second Lyapunov Method and Investigation of Stability of Some Resonance Problems, Dokl. Akad. Nauk SSSR, 193, No. 1, 46 - 49 (1970) (in Russian)

[44] HAPAYEV, M.M., Generalization of the Second Lyapunov Method and Investigation of Stability of Some Resonance Problems, IX Int. Conf. on Non - Linear Oscillations, Kiev, Naukova Dumka, (1981)

[45] HAPAYEV, M.M. AND BALANDIN, V.V., Construction of Perturbations of Lyapunov functions and Stability Investigations under Small Random Perturbations, Diff. Urav., 23, No 4, 675 - 680 (1987) (in Russian)

[46] HAPAYEV, M.M. AND FALIN, A.I., On Stability Investigation in Syatems Of Integro – Differential Equations by the Averaging Method. Dokl. Akad. Nauk SSSR, 250, No.2, 295 – 299 (1980) (in Russian)

[47] HAPAYEV, M.M. AND FILATOV, O.P., On Averaging and Stability in Systems with Singulariries, Dokl. Akad. Nauk SSSR, 261, No. 1, 67 – 70, (1981) (in Russian)

[48] HAPAYEV, M.M. AND FILATOV, O.P., On the Averaging Principle for Systems with Fast and Slow Variables, Diff. Urav., 19, No. 9, 1640 – 1643 (1983) (in Russian)

[49] HAPAYEV, M.M. AND FILATOV, O.P., Stability of Resonance Motions of Some Gyroscopic Systems, Solid state mech. No. 4, 69 – 72 (1982) (in Russian)

[50] HAPAYEV, M.M. AND KUZNETSOVA, I.V., Multi – Frequency Systems Containing Delay, Diff. Urav. 18, No. 2, 354 – 356 (1982) (in Russian)

[51] HAPAYEV, M.M. AND MAL'KOV, K.V., On Asymptotic Behaviour of Solutions of Perturbed Differential Equations, Dokl. Akad. Nauk SSSR, 290, No.4, 800 – 805 (1986) (in Russian)

[52] HAPAYEV, M.M. AND MAL'KOV, K.V., On a Class of Investigation Methods of Asymptotic Behaviour of Solutions of Differential Equations with Closed Operators, Diff. Urav., 22, No. 2, 255 – 267 (1986) (in Russian)

[53] HAPAYEV, M.M., Theorem of Lyapunov Type, Dokl. Akad. Nauk SSSR, 176, No. 6, 1262 – 1265 (1967) (in Russian)

[54] HAPAYEV, M.M., On Averaging Method in Some Problems Related to Averaging, Diff. Urav., 2, No. 5, 600 – 608 (1966) (in Russian)

[55] HAPAYEV, M.M., On Averaging in Multi – Frequency Systems, Dokl. Akad. Nauk SSSR, 217, No. 5, 1021 – 1024 (1974) (in Russian)

[56] HAPAYEV, M.M., On Evolution of Planetary Orbits, Dokl. Akad. Nauk SSSR, 312, No. 3, (1990) (in Russian)

[57] HAPAYEV, M.M., On Stability in the Three – Body Problem, Dokl. Akad. Nauk SSSR, 195, No. 2, 300 – 302 (1970) (in Russian)

[58] HAPAYEV, M.M., SHINKIN, V.N., On Investigation of Resonance Almost – Periodic Systems as Stability in Part of Variables, Prikl. Math. Mech., 47, No. 2, 334 – 337 (1983) (in Russian)

[59] HAPAYEV, M.M., On Stability Investigation in the Three – Body Problem Using a Hydrodynamics Model of Planet, Dokl. Akad. Nauk SSSR, 231, No. 5, 1092 – 1095 (1976) (in Russian)

[60] HAPAYEV, M.M., On Stability Investigation in the Theory of Non –Liner Oscillations, Mat. Zametki, 3, No. 3, 307 – 318 (1968) (in Russian)

[61] HAPAYEV, M.M., Stability of Equilibrium Position for Systems of Differential Equations, Diff. Urav., 5, No. 5, 848 - 855 (1969) (in Russian)

[62] HAPAYEV, M.M., Stability Problems in Systems of Ordinary Differential Equations, Usp. mat. nauk, 35, 1/211, 127 – 170 (1980) (in Russian)

[63] ILYNSHIN, A.A., LARIONOV, G.S., AND FILATOV, A.N., On Averaging in Systems of Non – Linear Integro – Differential Equations, Dokl. Akad. Nauk SSSR, 188, No. 1, 49 – 52 (1969) (in Russian)

[64] ISKANDER – ZADE, Z.A., Monotonic Stability of Motion in the case of Neutrality of Linear Approximation, Zh. Vichisl. Math. 6, No. 3, 454 – 465 (1966) (in Russian)

[65] KOLMOGOROV, A.N., On Conservation of Conditionally Periodic Motions under Small Variation of Hamiltonian Function, Dokl. Akad. Nauk SSSR, 98, No. 4, 527 – 530 (1954) (in Russian)

[66] KRASOVSKII, N.N., Some Problems of the Theory of Stability of Motion, Moscow, Fizmatgiz, 1959 (in Russian)

[67] KRYLOV, N.N., BOGOLYUBOV, M.N., New Methods in Linear Mechanics, Kiev, GTTs, (1934) (in Russian)

[68] KUNITSYN, A.L., MARKEYEV, A.P., Stability in Resonance Cases, In: Itogi Nauki i Tekhniki, Obshchaya Mekhanika, Moscow VINITI, 4, 58 – 139 (1979) (in Russian)

[69] KUZNETSOVA, I.V., On Averaging and Stability of Differential Equations with Deviating Argument. PhD Thesis, Moscow Univ. Press (1982) (in Russian)

[70] KUZNETSOVA, I.V. AND HAPAYEV, M.M., On Influence of Variable Tide Delay on Evolution of Orbits in System of Two Celestial Bodies, Astr. Zh., 61, No. 2, 371 – 374 (1984) (in Russian)

[71] KUZNETSOVA, I.V., On Averaging in Multi – Frequency Systems with Delay, Diff. Urav., 17, No., 6, 1128 – 1131 (1981) (in Russian)

[72] LYAPUNOV, A.M., General Problem of Motion Stability, Moscow – leningrad, Gostekhizdat, 1950 (in Russian)

[73] Lyapunov Vector Functions and their Construction, Novosibirsk Nauka (1980) (in Russian)

[74] MAC DONALD, G.J.F., Tidal Friction, Rev. geophys., 2, 467 – 511 (1964)

[75] MALKIN, I.G., Theory of Motion Stability, Moscow, Nauka (1966) (in Russian)

[76] MAL'KOV, K.V. AND HAPAYEV, M.M., On Damping Soliton - Like Solutions of Shallow Water Equations by Forestalling Resistance, Dokl. Akad. Nauk SSSR, 300, No. 5, 1052 – 1059 (1989) (in Russian)

[77] MAL'KOV, K.V. AND HAPAYEV, M.M., On Evolution and Stability Soliton - Like Solutions of the Boussinesque – Type Perturbed Equations, Dokl. Akad. Nauk SSSR, 292, No. 1, 68 – 73 (1987) (in Russian)

[78] MAL'KOV, K.V., On Construction of Stable and Stabilizing Systems with Disturbed Parameters, Diff. Urav., 25, No. 1, 74 – 89 (1989) (in Russian)

[79] MARKEYEV, A.P., Investigation of stability of Motion in Some Problems of Celestial Mechanics, Preprint, IPM SSSR, Moscow (1970) (in Russian)

[80] MARTYNENKO, YU.G., Motion of Unbalanced Gyroscope with Non – Contact Suspension, Mekh. Tverd. Tela, No. 4, 13 – 19 (1974) (in Russian)

[81] MARTYNENKO, YU.G. AND SAVCHENKO, T.A., Resonance Motion of Gyroscope with Non – Contact Suspension on Vibrating Base, Mekh. Tverd. Tela, No. 6, 16 – 24 (1977) (in Russian)

[82] MARTYNOV, D.YA., Close Double Stars and their Importance for the Theory of Star Evolution, Usp. Phys. nauk, 108, No. 4, 701 – 732, (1972) (in Russian)

[83] MARTYNOV, D.YA., Eclipsed Systems with Deformed Components: Finer Effects, in: Eclipsed variable Stars, Ed. V. P. Tsesevich, Moscow, Nauka, 1971 (in Russian)

[84] MASLOV, V.P., OMELYANOV, G.A., Asymptotic Soliton – Like Solutions of Equations with Small Dispersion, Usp. Math. Nauk, 36, No., 3, 63 – 67 (1981) (in Russian)

[85] MATROSOV, V.M., Motion Stability Theory, Prikl. Math. Mech., 26, No., 6, 992 – 1002 (1962) (in Russian)

[86] MERMAN, A., Instability of Some Canonical Systems in the Main Resonance Case, Bull. ITA SSSR, 14, No. 1, (154), 37 – 44 (1975) (in Russian)

[87] MITROPOLSKIY, YU. A., Averaging Method in Non – Linear Mechanics, Kiev, Naukova Dumka, 1971 (in Russian)

[88] MITROPOLSKIY, YU. A. AND FILATOV, A.N., Averaging of Integro – Differential and Integral Equations, Ukrain. Math. Zh., 24, No. 1, 30 – 40 (1972) (in Russian)

[89] MITROPOLSKIY, YU. A. AND MOSEYENKOV, B.I., Lectures on Applications of Asymptotic Methods to Solutions of Partial Differential Equations, Kiev, AN Ukrain. SSR Publishing House (1968) (in Russian)

[90] MOLCHANOV, A.M., Stability in the Case of Neutral Linear Approximation, Dokl. Akad. Nauk SSSR, 141, No. 1, 24 – 27 (1961) (in Russian)

[91] MOLCHANOV, A.M., The resonant Structure of the Solar System, Icarus, Int. J. of Solar System, 8, No. 2, 203 – 216 (1968)

[92] NEWELL, A.C., Solitons in Mathematics and Physics, Soc. for Ind. Appl. math.

[93] NOVIKOV, S.P., Periodic Problem for the Korteweg – de Vries Equation, Funk. Anal. Pril., 8 No. 3, 54 – 66 (1974) (in Russian)

[94] OZIRANER, A.S. RUMIANTSEV, V.V., The Lyapunov Function Method in the Problem of Stability of Motion in Part of Variables, Prikl. Math. Mech. 36, No. 2, 3634 – 384 (1972) (in Russian)

[95] POINCARÈ, H., Les Methodes Nouvelles de la Mechanique Celeste, Paris t. 1, (1892)

[96] POINCARÈ, H., Les Methodes Nouvelles de la Mechanique Celeste, Paris t. 2, (1893), t. 3 (1899)

[97] RUMIANTSEV, V.V., On Asymptotic Stability and Instability of Motion in Part of variables, Prikl. Math. Mech., 35, No. 1, 138 – 143 (1971) (in Russian)

[98] ROUCHE, N., HABETS, P., AND LALOY, H., Stability Theory by Lyapunov's Direct Method, Appl. math. Sci., 22, Springer Verlag, 1977

[99] SANSONE, G., Equazioi Differenziali nel Campo Real, Zanichelli, Bologna, 1948

[100] SHINKIN, V.N. On a Method of Investigating Stability by the Generalized Second Lyapunov Method by Means of Numerical Calculations,, Vestnik MGU, Ser 15, Comp. Math. and Cybernetics, No. 1, 36 – 43 (1980) (in Russian)

[101] SOKOLOV, V. YE., BABENKO, V.V., KOZLOV, L.F., ET. AL., State Discovery USSR, 1982, No. 265 (in Russian)

[102] Solitons and Nonlinear Wave Equations, Dodd, R.K. et. al. eds. Academic Press, 1982

[103] Strange Attractors. Kolmogorov, A.N. and Novikov, S.P. eds., Moscow, Mir, 1981 (in Russian)

[104] VALEYEV, K.G. AND DOLIA, V.V., On Dynamical Stabilization of Pendulum Oscillations, Prik. Math. Mech., 10, No. 2, 88 - 93 (1974) (in Russian)

[105] VENTZEL, A.D. AND FREIDLIN, M.I., Fluctuations in Dynamical Systems under Influence of Small Random Perturbations, Moscow (1979)

[106] VOLOSOV, V.M., Averaging in Systems of Ordinary Differential Equations Usp. Math. Nauk, 17, No. 6, 3 - 126 (1962) (in Russian)

[107] VOLOSOV, V.M. AND MORGUNOV, B.I., Averaging Method in Theory of Non - Linear Oscillating Systems, Moscow Univ. Press, (1971) (in Russian)

Index